Next-Level UI Development with PrimeNG

Master the versatile Angular component library to build stunning Angular applications

Dale Nguyen

Next-Level UI Development with PrimeNG

Group Product Manager: Rohit Rajkumar
Publishing Product Manager: Urvi Sambhav Shah
Senior Editor: Hayden Edwards
Technical Editor: Reenish Kulshrestha
Copy Editor: Safis Editing
Project Coordinator: Aishwarya Mohan
Proofreader: Safis Editing
Indexer: Manju Arasan
Production Designer: Ponraj Dhandapani
Marketing Coordinators: Nivedita Pandey and Anamika Singh

First published: March 2024

Production reference: 1010224

Published by Packt Publishing Ltd.
Grosvenor House
11 St Paul's Square
Birmingham
B3 1RB

ISBN 978-1-80324-981-0

`www.packtpub.com`

To Yen and Xoài, thanks for your great support. Having you both by my side has made a significant impact, and I am truly appreciative of the positive influence you've had on my life <3

- Dale Nguyen

Contributors

About the author

Dale Nguyen is a skilled full stack developer with a passion for technology. With years of experience in the industry, he has made significant contributions to agencies, education, finance, and travel sectors. Drawing from years of professional experience, Dale's expertise lies in Angular development. His dedication to staying updated with the latest advancements in Angular allows him to consistently deliver exceptional results. Proficient in multiple programming languages and tools, Dale is known for his ability to handle complex projects, with his advanced skills making him an invaluable asset to his employers and colleagues.

I want to thank the people who have supported me, especially my manager, Alex G.

About the reviewers

Alexandru Gogan is the VP of engineering at sherpa°. He started his career in public sector consultancy in Germany. His MBA in entrepreneurship from the University of Waterloo highlights his strategic approach, blending technical expertise with business insight. He was the first engineer at sherpa° and played a significant role in the company's growth. He specializes in web technologies, valuing a deep understanding of web fundamentals and infrastructure. As a leader, he is deeply committed to team development and cultural alignment, recognizing their importance in crafting a highly effective engineering team. His career reflects adaptability and a steadfast commitment to innovation in the tech industry.

Ajit Hingmire is a seasoned frontend developer who is highly skilled in Angular, TypeScript, PrimeNG, AG Grid, HTML5, CSS3, and optimizing web performance. He brings 15 years of technical experience in crafting responsive, high-performance applications and currently works as the AVP for a large European bank and works out of Pune, India. He is committed to elevating standards in code quality and UIs and adhering to best practices in frontend development. He is passionate about contributing insights to technical content, mentoring junior developers, and actively participating in technical meetups. In his free time, he likes to spend time with his son and play sports and games such as cricket, chess, and badminton.

Table of Contents

3

Utilizing Angular's Features and Improvements 27

4

Integrating PrimeNG into Your Angular Project 45

Part 2: UI Components and Features

5

Introducing Input Components and Form Controls 67

6

Working with Table, List, and Card Components 97

7

Working with Tree, TreeTable, and Timeline Components 131

8

Working with Navigation and Layout Components 157

Part 3: Advanced Techniques and Best Practices

9

Customizing PrimeNG Components with Theming 185

10

Exploring Optimization Techniques for Angular Applications 205

11

Creating Reusable and Extendable Components 231

12

Working with Internationalization and Localization 253

13

Testing PrimeNG Components 273

Part 4: Real-World Application

14

Building a Responsive Web Application 297

Preface

Angular is a powerful web application framework, and when combined with PrimeNG, a rich set of UI components, it becomes an even more potent tool for building cutting-edge web applications. This book provides a comprehensive guide to mastering PrimeNG in the context of Angular development.

Throughout the chapters, you will embark on a journey through the PrimeNG library, beginning with its integration into Angular projects and exploring its extensive set of UI components and features. Along the way, you will uncover advanced techniques and best practices, such as theming, performance optimization, and reusable components, culminating in real-world applications and case studies that demonstrate the power and versatility of PrimeNG.

Who this book is for

If you're an Angular developer or enthusiast eager to elevate your skills in crafting robust, visually appealing, and scalable web applications, then this book is for you. Whether you identify as a frontend developer, a full stack developer, or someone who places a premium on performance, you'll discover invaluable insights into customizing themes and seamlessly implementing responsive designs.

By reading this book, you will feel empowered to harness the full potential of PrimeNG, enabling you to create extraordinary web experiences that stand out from the rest.

What this book covers

In *Chapter 1*, *Introducing Angular and PrimeNG: A Powerful Combination*, you will be introduced to the powerful combination of Angular and PrimeNG for developing modern web applications. The chapter covers the basics of Angular and PrimeNG, their integration, and the advantages of using them together.

In *Chapter 2*, *Setting Up Your Development Environment*, you will be guided through setting up the development environment for building Angular applications with PrimeNG components. The chapter covers topics such as installing Node.js, the Angular CLI, and creating a new Angular project.

In *Chapter 3*, *Utilizing Angular's Features and Improvements*, you will explore the latest Angular features and their integration with PrimeNG components.

In *Chapter 4, Integrating PrimeNG into Your Angular Project*, you will see how to integrate PrimeNG into Angular projects, effectively combining the power of Angular and PrimeNG to create feature-rich applications. The chapter covers topics such as adding PrimeNG components, configuring PrimeNG modules, and customizing component styles and themes.

In *Chapter 5, Introducing Input Components and Form Controls*, you will see various input components and form controls provided by PrimeNG in their Angular applications. The chapter covers topics such as using text inputs, checkboxes, radio buttons, dropdowns, and more, as well as form validation and handling user input.

In *Chapter 6, Working with Table, List, and Card Components*, you will see PrimeNG data display components that present data effectively in Angular applications. The chapter covers topics such as using data tables, lists, and cards, as well as creating responsive layouts, handling data sorting, and implementing pagination.

In *Chapter 7, Working with Tree, TreeTable, and Timeline Components*, the focus will be on PrimeNG data presentation components that effectively manage data within Angular applications. The chapter covers topics such as working with tree structures, TreeTable, and Timeline, as well as handling user interactions and events.

In *Chapter 8, Working with Navigation and Layout Components*, you will be presented with PrimeNG navigation and layout components to create intuitive and user-friendly interfaces in their Angular applications. The chapter covers topics such as working with menus, breadcrumbs, tabs, and panels, as well as handling navigation events.

In *Chapter 9, Customizing PrimeNG Components with Theming*, you will see how to customize the appearance of PrimeNG components in their Angular applications using theming. The chapter covers topics such as working with pre-built themes, creating custom themes, using the Theme Designer, and overriding component styles.

In *Chapter 10, Exploring Optimization Techniques for Angular Applications*, you will discover tips and tricks for optimizing the performance of Angular applications that use PrimeNG components. The chapter covers topics such as lazy loading, change detection strategies, optimizing data binding, and using Angular's built-in performance tools.

In *Chapter 11, Creating Reusable and Extendable Components*, you will see how to create reusable and extendable components in Angular applications using PrimeNG. The chapter covers topics such as StyleClass, PrimeBlocks, creating reusable Angular components, and extending existing PrimeNG components.

In *Chapter 12, Working with Internationalization and Localization*, you will discover how to add internationalization and localization support to Angular applications using PrimeNG components.

In *Chapter 13, Testing PrimeNG Components*, you will receive guidance on testing Angular applications powered by PrimeNG components. The chapter also covers topics such as unit testing, using testing tools and libraries.

In *Chapter 14, Building a Responsive Web Application*, you will learn how to build a responsive web application using Angular and PrimeNG components. The chapter covers topics such as creating a project structure, implementing responsive layouts, integrating various PrimeNG components, and deploying the application.

To get the most out of this book

Before diving into this book, you should have a solid understanding of the fundamentals of Angular framework and web development concepts. Familiarity with HTML, CSS, and JavaScript, as well as TypeScript, is essential to get the most out of this book. Experience working with Angular components, directives, and services will also greatly enhance your learning journey.

While not required, basic knowledge of UI design principles and experience with other frontend libraries or frameworks can be beneficial in understanding the value PrimeNG brings to the Angular ecosystem. With these foundations in place, you'll be well-prepared to explore the powerful features and techniques covered in this comprehensive guide to PrimeNG mastery.

Software/hardware covered in the book	Operating system requirements
Angular 14+	Windows, macOS, or Linux
TypeScript 5+	
Node.js 18+	

The book is accompanied by a GitHub repository (a link is available in the next section) that contains code examples. If you encounter any difficulties at specific steps, please refer to the corresponding working version available on GitHub.

Download the example code files

You can download the example code files for this book from GitHub at `https://github.com/PacktPublishing/Next-Level-UI-Development-with-PrimeNG`. If there's an update to the code, it will be updated in the GitHub repository.

We also have other code bundles from our rich catalog of books and videos available at `https://github.com/PacktPublishing/`. Check them out!

Conventions used

There are a number of text conventions used throughout this book.

`Code in text`: Indicates code words in text, database table names, folder names, filenames, file extensions, pathnames, dummy URLs, user input, and Twitter handles. Here is an example: "These are just a few of the parameters available when running `ng new`. You can find more options and detailed explanations by running `ng new --help` or referring to the official Angular documentation for the specific version that you're using."

A block of code is set as follows:

```
<button (click)="handleClick()">Click me!</button>
...
handleClick() {
  // handle user click event
}
```

When we wish to draw your attention to a particular part of a code block, the relevant lines or items are set in bold:

```
export class UserListComponent {
  private userService = inject(UserService)
  users$ = this.userService.getUsers()
}
```

Any command-line input or output is written as follows:

```
nvm install 18
```

Bold: Indicates a new term, an important word, or words that you see onscreen. For instance, words in menus or dialog boxes appear in **bold**. Here is an example: "Select **System info** from the **Administration** panel."

> **Tips or important notes**
> Appear like this.

Get in touch

Feedback from our readers is always welcome.

General feedback: If you have questions about any aspect of this book, email us at `customercare@packtpub.com` and mention the book title in the subject of your message.

Errata: Although we have taken every care to ensure the accuracy of our content, mistakes do happen. If you have found a mistake in this book, we would be grateful if you would report this to us. Please visit `www.packtpub.com/support/errata` and fill in the form.

Piracy: If you come across any illegal copies of our works in any form on the internet, we would be grateful if you would provide us with the location address or website name. Please contact us at `copyright@packt.com` with a link to the material.

If you are interested in becoming an author: If there is a topic that you have expertise in and you are interested in either writing or contributing to a book, please visit `authors.packtpub.com`.

Share Your Thoughts

Once you've read *Next-Level UI Development with PrimeNG*, we'd love to hear your thoughts! Scan the QR code below to go straight to the Amazon review page for this book and share your feedback.

`https://packt.link/r/1-803-24981-1`

Your review is important to us and the tech community and will help us make sure we're delivering excellent quality content.

Download a free PDF copy of this book

Thanks for purchasing this book!

Do you like to read on the go but are unable to carry your print books everywhere?

Is your eBook purchase not compatible with the device of your choice?

Don't worry, now with every Packt book you get a DRM-free PDF version of that book at no cost.

Read anywhere, any place, on any device. Search, copy, and paste code from your favorite technical books directly into your application.

The perks don't stop there, you can get exclusive access to discounts, newsletters, and great free content in your inbox daily

Follow these simple steps to get the benefits:

1. Scan the QR code or visit the link below

https://packt.link/free-ebook/9781803249810

2. Submit your proof of purchase
3. That's it! We'll send your free PDF and other benefits to your email directly

Part 1: Introduction to PrimeNG

In the first part of this book, you will gain a comprehensive understanding of PrimeNG, a powerful UI component library for Angular. You will also gain knowledge on how to set up your development environment, ensuring it is properly prepared for integration. After that, you will learn how to leverage its extensive collection of components and features to enhance your application's user interface and functionality.

By the end of this part, you will be equipped with the knowledge and skills to effectively utilize PrimeNG in your Angular applications.

This part contains the following chapters:

1

Introducing Angular and PrimeNG: A Powerful Combination

Welcome to the first step of your journey into the vast and dynamic realm of web development!

This chapter serves as your gateway into two influential tools that have revolutionized the development of web applications: Angular and PrimeNG.

Angular, developed by tech giant Google, is a robust framework that has revolutionized the development of web applications. It provides a structured pathway for building complex and efficient web applications. With powerful features such as two-way data binding, dependency injection, and modular architecture, Angular enables developers to construct intricate web applications with ease.

In conjunction with Angular, PrimeNG offers a variety of pre-made UI components specifically designed for Angular. This combination forms a formidable pair, as PrimeNG complements Angular's capabilities by providing a suite of ready-to-use UI components when building all kinds of web applications. These components assist developers in crafting aesthetically pleasing and user-friendly interfaces, simplifying the development process and ensuring a consistent and engaging user experience.

In this chapter, we'll explore the fundamentals of both Angular and PrimeNG, demonstrating their integration and explaining the benefits of using them in conjunction. Whether you're a novice in the world of web development or a seasoned professional, this chapter promises to provide valuable insights that will pave the way for your future adventures in UI development.

To sum up, we're going to cover the following main topics:

- Introducing Angular
- Introducing PrimeNG
- Exploring key features of PrimeNG
- Using Angular and PrimeNG together

Technical requirements

We will be setting up the development environment in the next chapter; however, for now, here is where you can access the code for the book: `https://github.com/PacktPublishing/Next-Level-UI-Development-with-PrimeNG`. Note that you can find the code for each chapter in the `apps` folder.

Introducing Angular

Angular (`https://angular.io/`) is a web application framework developed and maintained by Google, and is a comprehensive framework for building dynamic web applications. As one of the pillars in the world of frontend development, Angular has established its dominance through its powerful features and performance-oriented nature.

When we dig into Angular, we can see that it's built with TypeScript, a statically typed superset of JavaScript. TypeScript brings in features such as type safety and enhanced tooling support, which contribute to the robustness of Angular applications. This combination of TypeScript's strong typing and Angular's architecture makes the framework highly scalable, a feature that's vital when it comes to building complex and large-scale applications.

Angular's architecture is component-based. A **component** controls a part of the screen on a website. Components are modular and reusable, which promotes a clean and **DRY (Don't Repeat Yourself)** code base. They also make it easier to manage and reason about your application as it grows in size and complexity.

You can see an example of a simple Angular component in the following code block:

```
// app.component.ts
import { Component } from '@angular/core'
import { CommonModule } from '@angular/common'
import { bootstrapApplication } from '@angular/platform-browser'
@Component({
  selector: 'my-app',
  standalone: true,
  imports: [CommonModule],
  template: `
    <h1>Hello from {{name}}!</h1>
    <a target="_blank" href="https://angular.io/start">
      Learn more about Angular
    </a>
  `,
})

// main.ts
export class AppComponent {
```

```
  name = 'Angular'
}
bootstrapApplication(AppComponent)
```

The code demonstrates a simple Angular application that uses the `bootstrapApplication()` function to initialize the application.

Let's break down the code and explain each part. First, we import the dependencies:

```
import { Component } from '@angular/core'
import { CommonModule } from '@angular/common'
import { bootstrapApplication } from '@angular/platform-browser'
```

Let's explain each part:

- The first line imports the `Component` decorator from `@angular/core`, which is used to define Angular components
- The second line imports `CommonModule` from `@angular/common`, which provides common directives, pipes, and services used by Angular applications
- The third line imports the `bootstrapApplication` function from `@angular/platform-browser`, which is used to bootstrap the Angular application

The next section is how we define an `Angular Component`:

```
@Component({
  selector: 'my-app',
  standalone: true,
  imports: [CommonModule],
  template: `
    <h1>Hello from {{name}}!</h1>
    <a target="_blank" href="https://angular.io/start">
      Learn more about Angular
    </a>
  `,
})
export class AppComponent {
  name = 'Angular'
}
```

Let's break down this code:

- This code defines a component named `AppComponent` using the `@Component` decorator
- The `selector` property sets the component's selector to `'my-app'`, which means it can be used in HTML with the `<my-app />` tag

- The `standalone` property is set to `true`, indicating that it can be used without being declared in an `NgModule`
- The `imports` array includes `CommonModule`, which provides common directives and services needed by the component

The `template` property defines the component's template, which consists of a `<h1>` heading displaying the value of the `name` property, `'Angular'`, and an anchor tag linking to the Angular website.

Finally, we bootstrap the application:

```
bootstrapApplication(AppComponent)
```

The `bootstrapApplication()` function is called with the `AppComponent` component as an argument, initiating an instance of an Angular application and rendering a standalone component as the application's root component.

> **Note**
>
> The standalone component approach (introduced in Angular v14) and the NgModule approach differ in how they are bootstrapped. With NgModule, bootstrapping is done using the `bootstrapModule()` function, which is exported from the `@angular/platform-browser-dynamic` package.

One of the strengths of Angular lies in its suite of powerful tools and features, including the following:

- The Angular **Command-Line Interface** (**CLI**) (`https://angular.io/cli`) simplifies the creation of Angular projects by automating various development tasks
- Features such as dependency injection and decorators enable developers to write code that's modular and easy to test
- Angular's directives allow developers to add behavior to DOM elements
- Two-way data binding keeps the model and the view in sync, reducing the amount of boilerplate code
- A comprehensive set of tools for handling complex state management and routing
- An HTTP client for interacting with RESTful services
- And more!

In summary, Angular is a well-rounded framework offering a combination of performance, scalability, and a vast array of tools and features. These factors make it a favorite among many developers, whether they're building a simple website or a complex single-page application.

> **Note**
>
> Angular is a constantly evolving framework, and new features are being added all the time. If you are interested in learning more about Angular, there are many resources available online (for example, `https://angular.io`).

In the next section, we will get into the world of PrimeNG.

Introducing PrimeNG

PrimeNG (`https://primeng.org/`) is a feature-rich library of open source UI components specifically designed for Angular applications. As you can see in the following figure, the current state of PrimeNG offers an impressive suite of 90+ components, 200+ icons, and 400+ ready-to-use UI blocks, ranging from simple widgets such as buttons and inputs to more complex and powerful components such as data tables, charts, and trees.

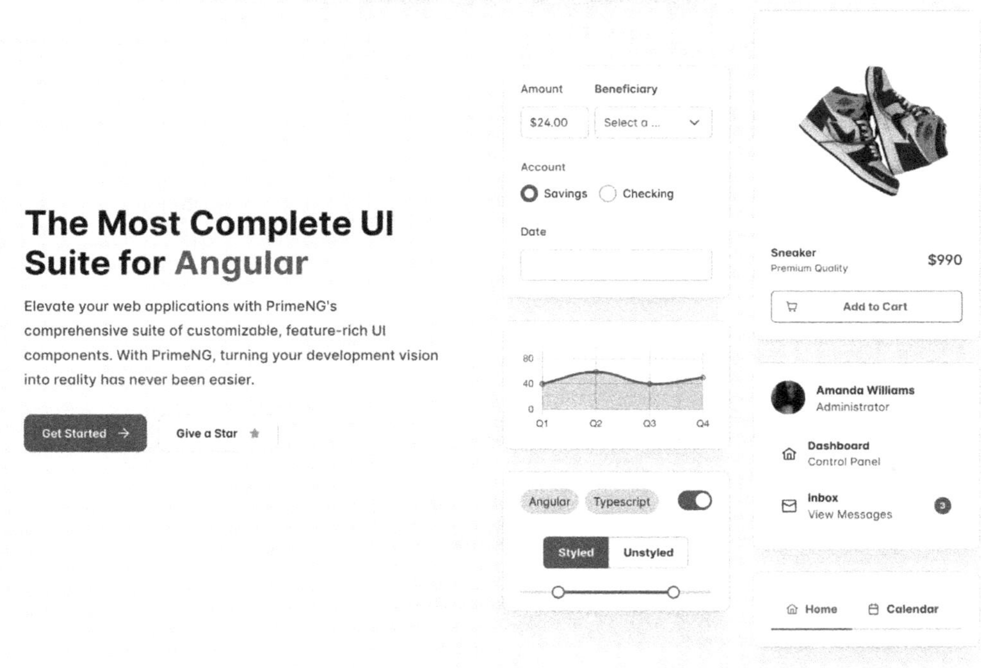

Figure 1.1 – Overview of PrimeNG

Every component in PrimeNG is crafted with attention to detail. They're not just functional but also aesthetically pleasing with a polished design that adheres to modern UI principles. The components come with a rich set of features out of the box, which can be further customized to suit the specific needs of your application.

Here is an example of a button snippet:

```
<section class="call-to-action">
   <h2>5 Things Your Spreadsheets Can't Do</h2>
   <button
      routerLink="/spreadsheets/demo"
      pButton
      pRipple
      label="Show Me Now"
      class="p-button-raised p-button-primary"
   ></button>
</section>
```

This code represents a `<button>` element with several attributes and classes. Let's break them down:

- `pButton`: This attribute is a directive that adds PrimeNG styling and behavior to the button. It enhances the button with additional features and visual enhancements.
- `pRipple`: This attribute enables the ripple effect on the button when clicked, providing visual feedback.
- `label="Show Me Now"`: This attribute sets the label or text content of the button to **Show Me Now**.
- `class="p-button-raised p-button-primary"`: This `class` attribute applies two CSS classes to the button – namely, `p-button-raised` and `p-button-primary`. These classes define the appearance and style of the button, making it raised and using a primary color scheme.

You can see the result in the following figure:

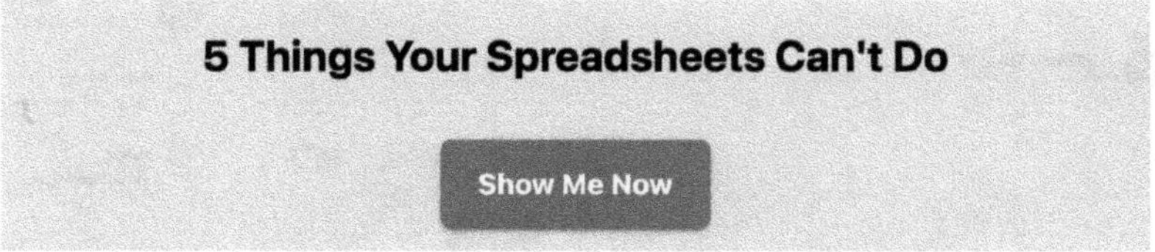

Figure 1.2 – PrimeNG button

In addition to its vast component library, PrimeNG is backed by a strong community and an active development team. Regular updates keep the library fresh and competitive, while the extensive documentation and examples make it easy to get started and find answers to your questions.

In conclusion, PrimeNG is a powerful tool that significantly reduces the time and effort required to build high-quality, interactive user interfaces in Angular applications. Its vast array of ready-to-use components, flexible theming capabilities, and commitment to accessibility make it an excellent choice for any Angular developer. In the next section, we will dive into the key features of PrimeNG.

> **Note**
>
> While PrimeNG provides a wealth of functionalities, an understanding of Angular basics is essential to leverage its full potential. You could visit the official Angular website or check out Packt's *Learning Angular* book by Aristeidis Bampakos and Pablo Deeleman.

Exploring key features of PrimeNG

PrimeNG stands out from other UI libraries with its rich feature set and attention to detail. This comprehensive set of features, designed specifically for Angular applications, makes it a go-to choice for developers looking for a powerful, easy-to-use UI library.

Let's check out some popular features of PrimeNG:

- One of the most noticeable features of PrimeNG is its wide array of UI components. From basic elements such as buttons and dropdowns to more complex components such as data tables, calendars, and charts, PrimeNG has you covered. Each component is fully customizable, giving you the flexibility to tailor the appearance and behavior to suit your needs.
- Another key feature of PrimeNG is its theming system. PrimeNG comes with a variety of pre-built themes, each with its own unique styling for all components, but the theming system doesn't stop there – with the help of the Theme Designer, you can easily create a custom theme that aligns with your brand identity. The Theme Designer provides a user-friendly interface for customizing the colors, fonts, and other styling aspects of your theme, as you can see here:

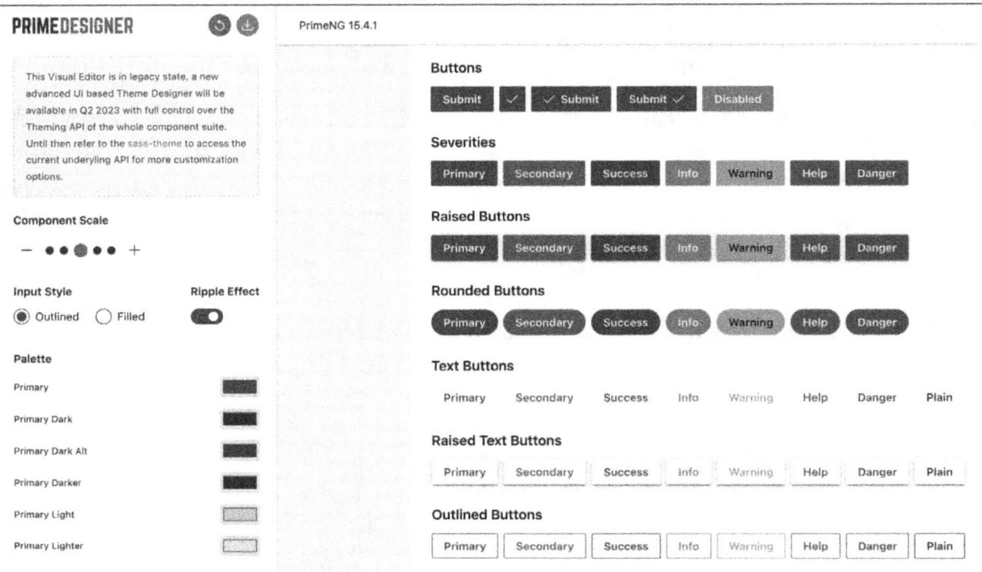

Figure 1.3 – PrimeNG Designer

- PrimeNG also prioritizes accessibility. Many of its components come with built-in support for **Accessible Rich Internet Applications** (**ARIAs**) and keyboard navigation, making your applications more accessible to users with disabilities. This commitment to accessibility is a testament to PrimeNG's dedication to creating inclusive web applications for people with disabilities. Using pre-built components with accessibility support saves time and effort by leveraging the expertise of accessibility specialists, ensuring adherence to standards, and benefiting from ongoing maintenance and updates.
- **PrimeIcons** is the default icon library of PrimeNG with over 250 open source icons developed by PrimeTek. These icons are not only visually appealing but also highly relevant in leveraging a common visual language of expected functionality and supporting the **user interface experience** (**UIX**). However, it is worth mentioning that PrimeNG components offer flexibility in using icons by allowing templating with other popular icon libraries such as Material Icons (`https://fonts.google.com/icons`) or Font Awesome (`https://fontawesome.com`), giving developers the freedom to choose the icon set that best aligns with their design requirements and desired user experience.
- Finally, PrimeNG is designed with responsiveness in mind. The components are built to adapt to different screen sizes and resolutions, ensuring your application looks great on all devices. Whether your users are on a desktop, tablet, or mobile device, PrimeNG components will provide a consistent, high-quality user experience.

In conclusion, PrimeNG offers a powerful set of features that make it an excellent choice for any Angular developer. From its extensive component library and flexible theming system to its commitment to accessibility and responsiveness, PrimeNG has all the tools you need to create high-quality, interactive user interfaces for your Angular applications.

Using Angular and PrimeNG together

The combination of Angular and PrimeNG offers a powerful set of tools for modern web development. Together, they create a highly productive environment that significantly simplifies the process of building complex, interactive web applications. Here are some highlights:

- First and foremost, PrimeNG is designed specifically for Angular. This means that all the components are built to work seamlessly with Angular's architecture. The PrimeNG components are essentially Angular components, so they can be used in the same way as any other Angular component. This compatibility with Angular results in a streamlined development process, where you can use PrimeNG components as part of your Angular application without any additional overhead or integration effort.
- Moreover, Angular and PrimeNG complement each other in terms of functionality. Angular provides the framework for building single-page applications with complex interactions and state management, while PrimeNG offers a wide array of pre-built UI components to enhance the user interface. This means you can leverage Angular's robust framework for the logic and structure of your application, and use PrimeNG's components to create an engaging, responsive user interface.
- The integration of Angular and PrimeNG also leads to improved code quality. With PrimeNG providing ready-made, reusable components, you can avoid repetitive code and focus on implementing unique features and business logic. This can lead to a cleaner, more maintainable code base, which is especially beneficial in large-scale applications.
- Finally, both Angular and PrimeNG have active communities and extensive documentation. This means you have a wealth of resources at your disposal when you need help or want to learn more about a specific topic. The consistent updates and improvements to both Angular and PrimeNG also ensure that you're working with up-to-date tools that embrace the latest best practices in web development.

In essence, the combination of Angular's powerful framework and PrimeNG's rich set of UI components provides a complete solution for building dynamic, aesthetically pleasing web applications. This blend not only enhances productivity but also leads to the development of applications that are efficient, scalable, and maintainable.

> **Note**
>
> Both Angular and PrimeNG are open source, ensuring they are not just free to use but that they can also be tailored to meet unique requirements such as custom components or theming.

Summary

In this chapter, we embarked on a journey into the world of Angular and PrimeNG, two robust tools that, when combined, unlock the potential to create modern and sophisticated web applications. Throughout this chapter, you gained an introduction to these tools and explored the advantages of leveraging them together. Angular provides the underlying structure, while PrimeNG enhances it with ready-to-use UI components. This blend enables developers to expedite their workflow, leading to quicker development cycles and more maintainable code.

The knowledge gained in this chapter is invaluable for professional developers seeking to enhance their web development skills. As we delve deeper into the subsequent chapters, we'll explore how to utilize Angular and PrimeNG to their full potential and create a diverse range of applications. Whether you're a budding developer or a seasoned professional, this journey promises to be insightful and enriching. So, fasten your seatbelts as we prepare to navigate the intriguing realm of web development with Angular and PrimeNG.

In the next chapter, we will delve into the process of setting up your development environment. We will explore the technical requirements, including Node.js, Yarn/NPM, GitHub, and VS Code, and guide you through the installation process. Additionally, we will introduce the Angular CLI, a powerful CLI that streamlines Angular development.

Get ready to take your development skills to the next level as we embark on this enlightening journey together!

2

Setting Up Your Development Environment

In this second chapter, we will dive into the crucial task of setting up your development environment to build Angular applications with PrimeNG components. This chapter equips you with the necessary knowledge and tools to create a seamless and productive development environment. From installing the required software to understanding the project structure, we will guide you through each step to ensure a smooth setup process.

By the end of this chapter, you will have a well-configured development environment and will be equipped with the necessary tools to start building Angular applications with PrimeNG. Understanding the technical requirements, setting up the Angular CLI, and familiarizing yourself with the project structure will lay a solid foundation for your web development journey. Additionally, leveraging an IDE such as VS Code and utilizing useful extensions will boost your productivity and make the development process more efficient.

So, let's dive in and set up your development environment for an optimal Angular development experience. In this chapter, we will cover the following topics:

- Setting up the Angular CLI
- Creating a new Angular project
- Understanding the project structure
- Discovering useful VS Code extensions

Technical requirements

The chapter contains various code samples from a new Angular project. You can find the related source code in the `chapter-02` folder of the following GitHub repository: `https://github.com/PacktPublishing/Next-Level-UI-Development-with-PrimeNG/tree/main/apps/chapter-02`.

Before diving into the setup process, it's important to ensure that your system meets the necessary technical requirements for development. Let's take a look at the key components you'll need to have in place:

- **Node.js (NVM)**: Install Node.js, a JavaScript runtime, using **Node Version Manager** (**NVM**) to manage multiple Node.js versions on your system. You can download and install NVM from the official website: `https://github.com/nvm-sh/nvm`. If your company imposes restrictions on the usage of NVM, please refer to the official Node.js website (`https://nodejs.org`) for installation instructions and follow the provided guide.
- **npm**: Choose npm for managing dependencies in your Angular projects. It comes bundled with Node.js, so if you have Node.js installed, you will be able to use npm. If you prefer to use alternatives to npm, you can check out either Yarn (`https://yarnpkg.com`) or pnpm (`https://pnpm.io`).
- **GitHub**: Sign up for a GitHub account to leverage this web-based hosting service for version control and collaboration. GitHub allows you to track changes, collaborate with team members, and host your Angular repositories. Sign up for an account at `https://github.com`.
- **VS Code (Visual Studio Code)**: Install VS Code, a free and extensible source-code editor developed by Microsoft. VS Code offers built-in support for Angular and integrates seamlessly with the Angular CLI, providing features such as code completion and debugging. Download VS Code from the official website: `https://code.visualstudio.com`.

By ensuring that you have Node.js (NVM), a package manager (npm, Yarn, or pnpm), GitHub, and VS Code installed, you'll have a solid foundation for setting up your Angular development environment. These tools will enable you to efficiently build, manage, and collaborate on your Angular projects with PrimeNG components.

Setting up the Angular CLI

The **Angular CLI** (**Command-Line Interface**) is a powerful tool that simplifies the process of creating, developing, and maintaining Angular applications. It provides a set of commands that automate common development tasks, allowing you to focus on building your application rather than setting up the project structure manually. In this section, we will guide you through the installation process of the Angular CLI and provide an overview of its core commands.

> **Note**
>
> Ensure that you have a stable internet connection during the installation process. It may take some time to download and install the required packages, depending on your internet speed.

Installing Node.js v18 (using NVM)

To install NVM and set Node.js v18 as the default version, follow these steps:

1. Visit the official NVM repository on GitHub: `https://github.com/nvm-sh/nvm`.
2. Follow the installation instructions specific to your operating system. This typically involves running a script to download and install NVM. The following script will help to download and install NVM v0.39.3:

   ```
   curl -o- https://raw.githubusercontent.com/nvm-sh/nvm/v0.39.3/
   install.sh | bash
   ```

> **Note**
>
> The version of NVM may change, so please visit the official website for the latest version and instructions.

3. Open a new terminal window or restart your terminal to load NVM.
4. Run the following command to verify that NVM is installed:

   ```
   nvm --version
   // result
   0.39.3
   ```

   You should see the version number of NVM printed in the terminal. In this example, the current version is `0.39.3`.

5. After that, run the following command to install Node.js v18 using NVM:

   ```
   nvm install 18
   ```

6. In order to use Node.js v18 in a new terminal session, we need to set Node.js v18 as the default version. To do that, run the following command:

   ```
   nvm alias default 18
   ```

7. Finally, run the following commands to verify that Node.js v18 is installed and set as the default:

   ```
   node --version
   npm --version
   ```

These commands should display the version numbers of Node.js and npm respectively, and they should correspond to the installed Node.js v18.

With Node.js v18 installed, let's move on to installing the Angular CLI.

Installing the Angular CLI

To install the Angular CLI, complete the following instructions depending on your operating system.

For Windows machines, do the following:

1. Open Command Prompt or PowerShell.
2. Run the following command to install the Angular CLI globally:

```
npm install -g @angular/cli
```

For Linux/macOS machines, do the following:

1. Open Terminal.
2. Run the following command to install the Angular CLI globally:

```
npm install -g @angular/cli
```

> **Note**
> If you encounter permission errors while installing Node.js or the Angular CLI using the npm package manager, you may need to use `sudo` before the commands to run them with administrator privileges.

For macOS (using Homebrew), do the following:

1. Open Terminal.
2. Install Homebrew by running the following command:

```
/bin/bash -c "$(curl -fsSL https://raw.githubusercontent.com/
Homebrew/install/HEAD/install.sh)
```

3. Once Homebrew is installed, run the following command to install Node.js:

```
brew install angular-cli
```

After following the appropriate installation process for your system, confirm the version of the Angular CLI with this command:

```
ng version
```

You will receive the output shown in *Figure 2.1*. As you can see, at the time of writing this book, the version of the Angular CLI is `17.0.6`.

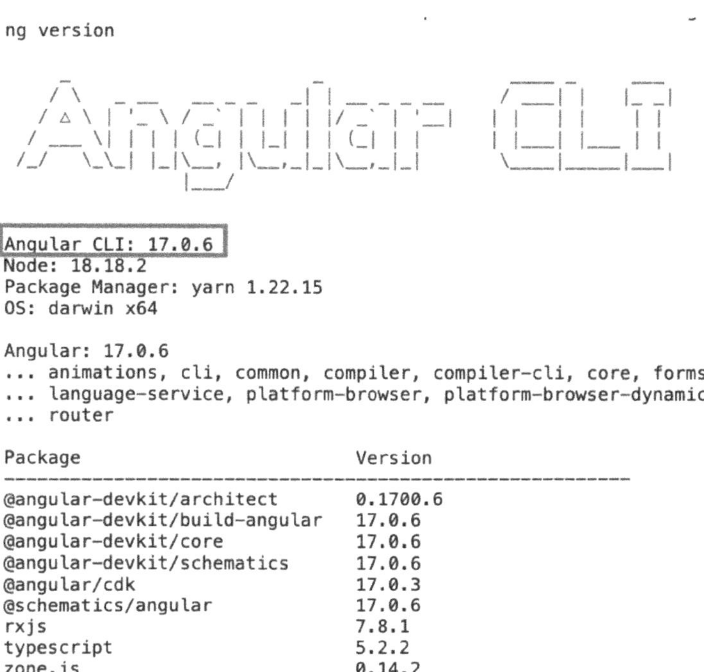

Figure 2.1 – The Angular CLI version

The Angular CLI provides a wide range of commands to streamline the development process. Here are some of the most commonly used commands and their explanations:

- `ng new [project-name]`: Creates a new Angular project with the specified name. It sets up the project structure, installs dependencies, and generates initial boilerplate code.
- `ng serve`: Starts a development server and compiles your Angular application. It watches for changes in your files and automatically reloads the application in the browser.
- `ng generate [schematic] [name]`: Generates different elements of your Angular application such as components, services, modules, and more. It scaffolds the necessary files and updates the required configurations.
- `ng build`: Builds the Angular application for production. It compiles the code and generates optimized files that can be deployed to a web server.
- `ng test`: Executes unit tests for your Angular application. It runs the tests using the configured test runner and provides detailed information about test results.
- `ng lint`: Analyzes the code for potential errors and code style violations. It helps enforce coding standards and maintain code quality.

- `ng deploy`: Deploys your Angular application to a hosting platform, such as GitHub Pages or Firebase Hosting. It automates the deployment process and makes your application accessible to the public.

> **Note**
> You can run `ng help` to see the list of commands and their usage. You can also check the official documentation for an overview of all commands at `https://angular.io/cli#command-overview`.

By leveraging the Angular CLI, you can streamline your development workflow, automate repetitive tasks, and focus on building high-quality Angular applications. The installation process on Windows, Linux, and macOS, including an alternative option using Homebrew for macOS, ensures that you have the necessary tools to harness the power of the Angular CLI.

Now that you have the Angular CLI set up, explore the various commands it offers to create, build, test, and deploy your Angular applications. With the Angular CLI, you'll enhance your productivity as a professional developer and unlock the full potential of Angular for building robust and scalable web applications. In the next section, we will start to create a new Angular project.

Creating a new Angular project

Creating a new Angular project is a straightforward process. In this section, we will guide you through the steps of creating a new Angular project. We will also explore the structure and purpose of each file in a new Angular project.

To create a new Angular project, follow these steps:

1. Open your Command Prompt or Terminal.
2. Navigate to the directory where you want to create your project.
3. Run the following command to generate a new Angular project, replacing `my-app` with the desired name of your project:

   ```
   ng new my-app
   ```

   The `ng new` command creates a new Angular project with the default configuration and project structure. It installs the necessary dependencies and sets up the initial files for your application.

When running this command to create a new Angular project, there are several parameters (flags) you can use to customize the project setup. Here are some commonly used parameters:

- `--dry-run`: Performs a dry run of the project generation without actually creating the files. It allows you to see the files that would be generated before committing to the project creation.
- `--standalone`: Creates an application based upon the standalone API, without NgModules.

- `--inline-style` or `--inline-template`: Specifies whether to use inline styles or templates. By default, Angular generates separate style and template files. Using these flags, you can choose to have inline styles or templates within the component files.
- `--prefix`: Sets the prefix for the generated component selector. The prefix is added to the selector of every component generated in the project.
- `--style`: Specifies the style format to use in the project, such as CSS, SCSS, Sass, Less, or Stylus. For example, `--style=scss` will configure the project to use SCSS as the default style format.
- `--routing`: Generates a routing configuration for the initial project.
- `--skip-git`: Skips initializing a new Git repository in the project directory. This is useful if you prefer to manage version control manually.
- `--skip-tests`: Prevents the generation of spec files for unit tests when creating new components. Use this flag if you don't want test files to be generated by default.
- `--skip-install`: Skips the installation of npm packages after project creation. Use this if you prefer to manually run `npm install` or `yarn` later to install dependencies.
- `--directory`: Specifies the directory name to create the project in. By default, the project is created in a folder with the same name as the project.
- `--minimal`: Create a workspace without any testing frameworks (use for learning purposes only).

Here is the command to generate a new Angular project with such options:

```
ng new primeng-demo --standalone --style=scss --inline-style --inline-
template
```

> **Note**
>
> Using the `--standalone` option is highly recommended as it reduces boilerplate code and became the default behavior in Angular 17.

These are just a few of the parameters available when running `ng new`. You can find more options and detailed explanations by running `ng new --help` or referring to the official Angular documentation for the specific version that you're using.

After running the script, the Angular CLI will ask to enable Server-Side Rendering (SSR) and Static Site Generation (SSG/Prerendering). Select `NO` as the answer, since it's not relevant at this moment:

```
? Do you want to enable Server-Side Rendering (SSR) and Static Site
Generation
(SSG/Prerendering)? (y/N)N
```

Figure 2.2 shows the final result of creating your new Angular project:

```
CREATE primeng-demo/README.md (1065 bytes)
CREATE primeng-demo/.editorconfig (274 bytes)
CREATE primeng-demo/.gitignore (548 bytes)
CREATE primeng-demo/angular.json (3160 bytes)
CREATE primeng-demo/package.json (1043 bytes)
CREATE primeng-demo/tsconfig.json (901 bytes)
CREATE primeng-demo/tsconfig.app.json (263 bytes)
CREATE primeng-demo/tsconfig.spec.json (273 bytes)
CREATE primeng-demo/.vscode/extensions.json (130 bytes)
CREATE primeng-demo/.vscode/launch.json (470 bytes)
CREATE primeng-demo/.vscode/tasks.json (938 bytes)
CREATE primeng-demo/src/main.ts (250 bytes)
CREATE primeng-demo/src/favicon.ico (948 bytes)
CREATE primeng-demo/src/index.html (297 bytes)
CREATE primeng-demo/src/styles.scss (80 bytes)
CREATE primeng-demo/src/app/app.component.spec.ts (906 bytes)
CREATE primeng-demo/src/app/app.component.ts (1552 bytes)
CREATE primeng-demo/src/app/app.config.ts (117 bytes)
CREATE primeng-demo/src/assets/.gitkeep (0 bytes)
✔ Packages installed successfully.
    Successfully initialized git.
...
```

Figure 2.2 – The Angular CLI result

After that, you can run `ng serve` to check the newly created project, like so:

```
> ng serve

Initial Chunk Files | Names          | Raw Size
polyfills.js        | polyfills      | 82.71 kB |
main.js             | main           |  1.57 kB |
styles.css          | styles         | 96 bytes |

                    | Initial Total | 84.38 kB
Application bundle generation complete. [2.899 seconds]
Watch mode enabled. Watching for file changes...
      Local:    http://localhost:4200/
```

Then you can visit `http://localhost:4200/` to check your web app – see *Figure 2.3*:

Welcome to primeng-demo!

Figure 2.3 – Angular demo app

Now that we have our first Angular application created, let's go through its structure.

Understanding the project structure

Understanding the purpose of each file in your Angular project is essential for navigating and developing your application effectively. Each file plays a specific role in the overall structure and functionality of your Angular project.

The following is a brief overview of the new structure of Angular:

- **The root directory**: The root directory refers to the main folder that contains all the project files and directories. It contains the following files:
  - `README.md`: Contains a description of the Angular application.
  - `.editorconfig`: Contains configuration for code editors.
  - `.gitignore`: Specifies intentionally untracked files that Git should ignore.
  - `angular.json`: Contains CLI configuration defaults for all projects in the workspace, including configuration options for build, serve, and test tools that the CLI uses.
  - `package.json`: Specifies the application's dependencies, devDependencies, scripts, licensing, and so on.
  - `tsconfig.json`: Specifies the TypeScript compiler configuration for the Angular application.
  - `tsconfig.app.json`: Specifies the TypeScript compiler configuration for the application's main module.
  - `tsconfig.spec.json`: Specifies the TypeScript compiler configuration for the application's unit tests.
- **The src directory**: The `src` directory contains the source code for the Angular application. It is divided into the following subdirectories:
  - `main.ts`: The entry point for the Angular application.
  - `favicon.ico`: The application's favicon.
  - `index.html`: The application's main HTML file.
  - `styles.scss`: The application's main CSS file.
  - app: Contains the application's components, services, directives, pipes, and so on.
  - `app.component.spec.ts`: The unit test for the application's main component.
  - `app.component.ts`: The definition of the application's main component.
  - `app.config.ts`: The configuration file for the application's main entry point.
  - `assets`: Contains the application's assets, such as images and fonts.

The new structure of Angular is a significant improvement over the previous structure (before Angular 14). It makes it easier to develop and maintain Angular applications, and it is more consistent with the way other web frameworks are structured.

> **Note**
>
> Angular has released a new site which provides more tutorials and lessons on the latest Angular features. You can learn more at `https://angular.dev`.

Now that you have created your Angular project and explored the project structure, you are ready to start building your application by leveraging the power of Angular and PrimeNG components. Before starting, let's go through some useful VS Code extensions that will help us during the development process.

Discovering useful VS Code extensions

When it comes to developing Angular applications, having the right tools can greatly enhance your productivity and efficiency. One of the most popular code editors among developers is **Visual Studio Code (VS Code)**. VS Code has a wide range of extensions that can help streamline your Angular development workflow. In this section, we will introduce some useful VS Code extensions specifically tailored for Angular development.

Angular Language Service

The **Angular Language Service** extension is an invaluable tool for Angular developers. This extension provides a rich editing experience for Angular templates, both inline and external, including the following:

- **Completion lists**: Provides suggestions and autocompletion for Angular template syntax, helping developers write code more efficiently and accurately
- **AOT diagnostic messages**: Displays compile-time diagnostic messages related to **Ahead-of-Time (AOT)** compilation in Angular templates, helping developers catch errors and improve code quality
- **Quick info**: Provides contextual information and documentation about Angular directives and components when hovering over them in the template, aiding developers in understanding and using the API effectively
- **Go to definitions**: Allows developers to navigate to the definition of a symbol in the template, making it easier to understand how components and directives are implemented and facilitating code exploration and debugging

Figure 2.4 shows an example of the autocomplete feature from the extension – here we type `heading` in the template and the extension gives the autocomplete option to the property from the component:

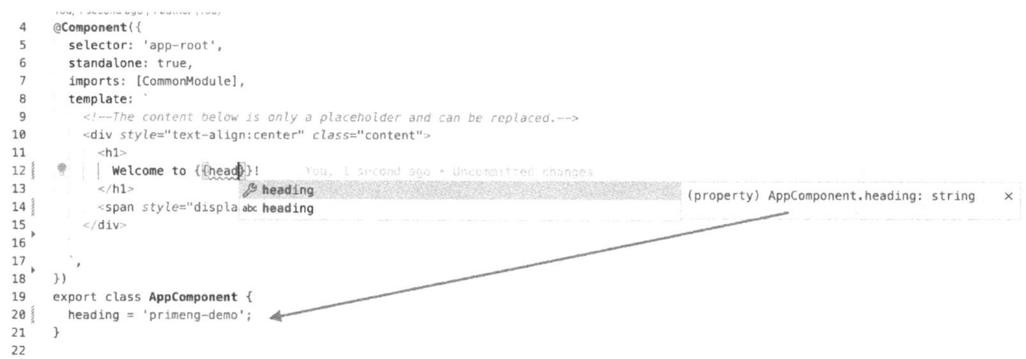

```
@Component({
  selector: 'app-root',
  standalone: true,
  imports: [CommonModule],
  template: `
    <!--The content below is only a placeholder and can be replaced.-->
    <div style="text-align:center" class="content">
      <h1>
        Welcome to {{head}}!
      </h1>
      <span style="displa
    </div>

  `,
})
export class AppComponent {
  heading = 'primeng-demo';
}
```

heading
heading
(property) AppComponent.heading: string

Figure 2.4 – Angular Language Service autocomplete example

> **Note**
>
> Template autocomplete only works with public properties from the component.

Editor Config

The **Editor Config** extension ensures consistent coding styles across your development team. It reads the `.editorconfig` file in your project and applies the defined rules to your code. With Editor Config, you can enforce indentation styles, line endings, encoding, and other formatting preferences. This extension is especially useful when collaborating with other developers on Angular projects, as it helps maintain a unified code style and minimizes style-related conflicts.

Here is an example of the `.editorconfig` file in a newly created Angular project:

```
# Editor configuration, see https://editorconfig.org
root = true

[*]
charset = utf-8
indent_style = space
indent_size = 2
insert_final_newline = true
trim_trailing_whitespace = true

[*.ts]
quote_type = single

[*.md]
max_line_length = off
```

```
trim_trailing_whitespace = false
```

As you can see, the rules are descriptive – for example, setting the character encoding of the files to UTF-8 or using single quotes for all TypeScript files.

Angular Schematics

Angular Schematics is a powerful extension that integrates with the Angular CLI and provides a scaffolding mechanism for generating and modifying code. It allows you to generate components, modules, services, and other Angular artifacts with ease. With Angular Schematics, you can quickly create boilerplate code and follow consistent patterns and practices within your Angular project. It saves time by automating repetitive tasks and helps maintain a standardized structure across your code base.

You can see in *Figure 2.5* that there is a list of options, and we can generate a component named `about` without remembering the command detail:

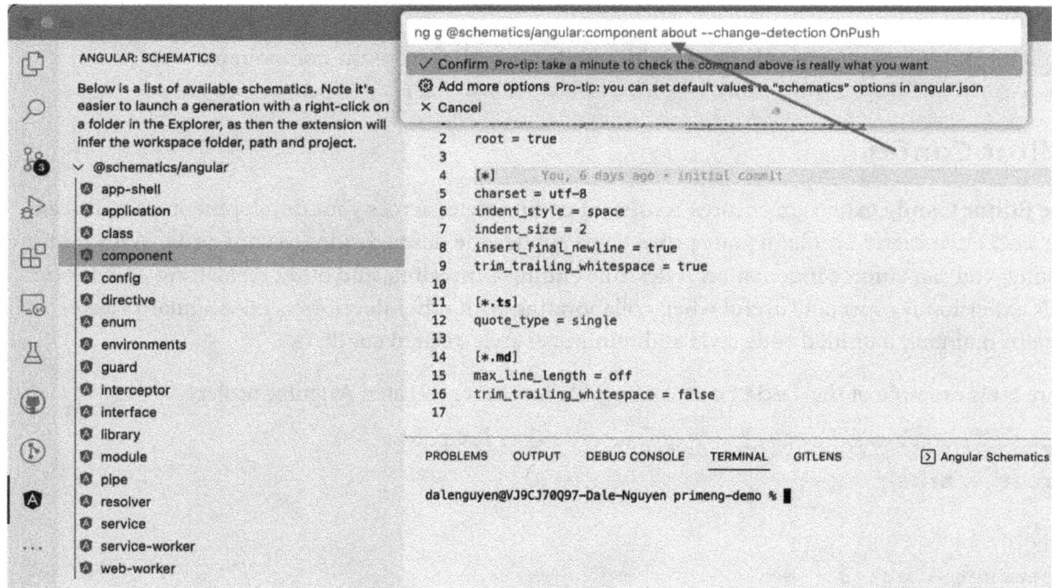

Figure 2.5 – Angular Schematics

Auto Rename Tag

The **Auto Rename Tag** extension is a time-saving tool that automatically renames HTML tags as you edit them. When you modify the opening or closing tag of an element, this extension updates the corresponding tag throughout your code base, ensuring consistency and preventing tag mismatches.

It eliminates the need for manual tag renaming, which can be error-prone and time-consuming, especially in larger Angular projects with complex HTML structures.

Nx Console

This is a bonus section. If you prefer using **Nx Workspace** (`https://nx.dev`) for Angular development, this extension is for you. The **Nx Console** extension for VS Code streamlines Angular development, offering code generation, dependency graph visualization, and productivity-enhancing features directly within the IDE. It boosts efficiency, enhances code quality, and accelerates development workflows.

Summary

This chapter is dedicated to the crucial task of setting up your development environment to build Angular applications with PrimeNG components. We began by discussing the technical requirements, which include Node.js, Yarn/npm, GitHub, and VS Code. Detailed installation instructions were provided for each of these tools, ensuring that you have the necessary prerequisites for a seamless development experience.

The chapter then focused on the Angular CLI, a powerful command-line interface for Angular development. We walked you through the installation process on Windows, Linux, and macOS, enabling you to leverage the Angular CLI's extensive functionality, including scaffolding, building, and testing Angular applications. Additionally, we covered creating a new Angular project, exploring the latest standalone component option and the Angular project template. You should now have a well-configured development environment equipped with the Angular CLI, ready to embark on the exciting journey of building modern web applications with Angular and PrimeNG.

Furthermore, we introduced you to several indispensable VS Code extensions tailored specifically for Angular development. These extensions, such as Angular Language Service, Editor Config, Angular Schematics, and Auto Rename Tag, significantly enhance your coding experience and boost productivity. With features such as intelligent code completion, formatting assistance, and workspace management, these extensions ensure that you maintain consistent coding styles and streamline your development workflow within the VS Code environment. By harnessing the power of these extensions, you'll be well equipped to tackle the challenges of Angular development and maximize your efficiency throughout the development process.

With your development environment fully set up and optimized, you are now primed to dive into the subsequent chapters and explore the full potential of Angular and PrimeNG in creating exceptional web applications. Specifically, in the next chapter, we'll go over Angular's core features and improvements released in the latest version.

3

Utilizing Angular's Features and Improvements

Welcome to the exciting world of modern Angular development. With its rapid six-month release cycle, Angular has evolved dramatically, introducing new features and updates that change how we approach our applications in a really interesting way!

In this chapter, we will explore Angular 17, the latest version of the Angular framework at the time of writing, and discover how its new features and improvements empower developers to build cutting-edge web applications. We will also delve into the core features of Angular and understand how they enhance the development process.

In this chapter, we will cover the following topics:

- Introducing modern Angular
- Learning about the core features and improvements in Angular
- Organizing an Angular project

Technical requirements

This chapter contains various code samples of Angular core features and concepts. You can find the related source code in the `chapter-03` folder of the following GitHub repository: `https://github.com/PacktPublishing/Next-Level-UI-Development-with-PrimeNG/tree/main/apps/chapter-03`.

Introducing modern Angular

As we embark on this chapter, let's take a moment to explore the Angular framework's journey and understand how it has evolved over time to become a powerhouse in web development. From its inception to the latest version, Angular has continuously evolved to meet the changing needs of developers and provide a robust foundation for building complex web applications.

The Angular framework was initially introduced by Google in 2010 as AngularJS (using JavaScript). It was a game-changer in the web development landscape, offering a declarative and powerful approach to building dynamic user interfaces. However, as web technologies advanced and developers demanded more scalability and performance, Angular underwent a significant transformation.

One of the pivotal moments in Angular's evolution was the release of Angular 2 (using TypeScript) in 2016. Angular 2 introduced a complete rewrite of the framework, embracing modern concepts such as component-based architecture and reactive programming. This shift laid the groundwork for Angular's future growth and set the stage for subsequent versions.

In subsequent releases, Angular continued to refine its features and introduce new capabilities such as bundle-size optimization, backward compatibility, animation, tree-shaking, and server-side rendering. Notably, Angular version 14 brought significant improvements to the framework. It emphasized a more modular approach with the introduction of standalone components, enabling developers to create reusable and encapsulated code. This modular architecture revolutionized the way developers approached Angular development, enhancing code maintainability and reusability.

Understanding these changes allows you to leverage the latest features and best practices, leading to more efficient and scalable applications. In the next section, let's have an overview of the existing features and new improvements introduced in recent versions of Angular.

Learning about the core features and improvements in Angular

In this section, let's explore some of the features and improvements that Angular brings to the table, including data binding, components, services, directives, pipes, signals, and control flow. It's important to understand some core concepts of Angular, so we can have a better understanding of how Angular and PrimeNG work together in the following chapters.

Angular data binding

Data binding is a fundamental concept in Angular that enables the synchronization of data between the component and the view. It allows you to establish a connection between the data in your component and the HTML elements in the template. Data binding ensures that any changes in the component are automatically reflected in the view, and vice versa.

Angular supports several types of data binding. Let's explore each type:

- **Interpolation**: Interpolation is a common way to display data from a component to the UI. It is a one-way data binding mechanism that allows you to embed dynamic values within the template. It uses double curly braces, `{{ }}`, to bind the component property to the view. Here's an example:

  ```
  <p>Welcome, {{ username }}!</p>
  ```

In the code, the `username` property from the component is interpolated into the `<p>` element, as follows:

Figure 3.1 – Interpolation example

- **Property binding**: Property binding allows you to establish a one-way binding of a component value to an HTML element property. It is denoted by square brackets, `[ ]`. Here's an example:

  ```
  <input [value]="username" />
  ```

  In the previous code, the value of the `<input>` element is bound to the `username` property from the component.

- **Event binding**: Event binding allows you to listen to events triggered by the user and perform actions in response to those events. It establishes a one-way binding from the view to the component. Event binding is denoted by parentheses, `()`. Here's an example:

  ```
  <button (click)="handleClick()">Click me!</button>
  ...
  handleClick() {
    // handle user click event
  }
  ```

  In the code, the `handleClick()` method from the component is bound to the `click` event of the `<button>` element.

- **Two-way binding**: Two-way binding combines both property binding and event binding, enabling synchronization of data in both directions. It allows changes in the view to update the component and vice versa. Two-way binding is denoted by the `[(ngModel)]` directive, which synchronizes the value of a form control in the template with a property in the component class. Here's an example:

```
// FormsModule needs to be imported for this work on form input
import { FormsModule } from '@angular/forms'

<input [(ngModel)]="username" />
```

In the code, the value of the `<input>` element is bound to the `username` property from the component. Any changes in the input field will update the component property, and vice versa.

To use data binding in Angular, you need to define the component property and bind it to the appropriate HTML element in the template. Let's look at an example of using interpolation and property binding:

```
@Component({
  selector: 'app-greeting',
  template: `
    <p>Welcome, {{ username }}!</p>
    <input [value]="username" />
  `
})
export class GreetingComponent {
  username = 'John Doe';
}
```

Here, `GreetingComponent` has a `username` property set to `'John Doe'`. The `{{ username }}` instance in the `<p>` element uses interpolation to display the value of `username` in the template. The `[value]="username"` instance in the `<input>` element uses property binding to set the initial value of the input field.

Angular components

When building applications with Angular, components play a central role in defining the user interface and behavior of different parts of the application. A **component** in Angular encapsulates the HTML template, CSS styles, and TypeScript code required to render a specific part of the user interface. It promotes reusability, maintainability, and modularity, by dividing the application into smaller, self-contained units.

In previous versions of Angular, components were typically registered and managed within Angular modules using the `@NgModule` decorator. The `@NgModule` decorator was used to define a module, which is a container for related components, directives, services, and other artifacts. Within `@NgModule`, you could specify the components that belong to it using the `declarations` property. Here's an example:

```
@NgModule({
  declarations: [AppComponent, HeaderComponent, FooterComponent],
  // Other module properties...
})
export class AppModule { }
```

In this code snippet, the `declarations` property includes three components: `AppComponent`, `HeaderComponent`, and `FooterComponent`. These components can be used within the module or exported to be used in other modules.

Since Angular v14, we have had another concept, which is standalone components. A **standalone component** is a component that lives on its own without being listed in the `declarations` array under `@NgModule`. Let's create a new standalone component from scratch.

Let's use Angular CLI to generate a new standalone component:

```
ng g c alert
...
CREATE src/app/alert/alert.component.scss (0 bytes)
CREATE src/app/alert/alert.component.html (20 bytes)
CREATE src/app/alert/alert.component.spec.ts (547 bytes)
CREATE src/app/alert/alert.component.ts (294 bytes)
```

> **Note**
>
> If you are using an Angular version lower than v17, make sure to append `--standalone` when generating standalone components.

The command successfully generated the alert component in the specified directory, along with its associated files for styles, template, test, and component:

- `alert.component.scss`: This file is intended for defining styles specific to the alert component
- `alert.component.html`: This file is meant to contain the template code for the alert component
- `alert.component.spec.ts`: This file is used for writing tests for the alert component
- `alert.component.ts`: This file is the main file where the logic and behavior of the alert component will be implemented

> **Note**
>
> You can further enhance your application's size and maintainability by using the `--inline-style` and `--inline-template` options, enabling you to incorporate inline styles and templates. This approach reduces file dependencies, resulting in a more streamlined and manageable application. This approach will also encourage you to write smaller and more maintainable components since a big anti-pattern is to have singular complex and huge components.

Let's now look at the alert component in detail:

```
import { Component } from '@angular/core';
import { CommonModule } from '@angular/common';

@Component({
  selector: 'app-alert',
  standalone: true,
  imports: [CommonModule],
  templateUrl: './alert.component.html',
  styleUrls: ['./alert.component.scss']
})
export class AlertComponent {}
```

The provided code defines an `AlertComponent` instance within an Angular application. Let's break down the code:

- `@Component({ ... })`: This is the decorator used to define an Angular component. It provides metadata about the component.
- `selector: 'app-alert'`: This property specifies the HTML tag selector for this component. In this case, the component can be used in templates with the `<app-alert />` tag.
- `standalone: true`: This indicates that the component is self-contained and can be used by importing to another standalone component or `NgModule`.
- `imports: [CommonModule]`: This property specifies the modules or components that should be imported for this component. In this case, it imports the `CommonModule`, which is required for using common Angular directives.
- `templateUrl: './alert.component.html'`: This property specifies the URL of the external HTML template file associated with this component.
- `styleUrls: ['./alert.component.scss']`: This property specifies an array of external stylesheet URLs associated with this component.

It's recommended that when starting a new project, we should utilize standalone components since it will help simplify the way we build Angular applications. It's also better for future migration since standalone components will be the default option when creating a new application in Angular 17.

> **Note**
>
> Angular still supports `NgModule`, and you can also mix `NgModule` and standalone components without any issues. More than that, we can also create standalone directives or pipes.

In summary, Angular components are the building blocks of Angular applications, providing encapsulated functionality and UI representation. By leveraging the power of components, you can create reusable and modular code that promotes maintainability and code organization. Now, let's explore another fundamental concept in Angular development: dependency injection.

Dependency injection

At its core, **dependency injection** (**DI**) is a design pattern that allows objects to receive their dependencies from an external source rather than creating them internally. In Angular, the DI system takes care of providing the required dependencies to the components, services, and other Angular constructs. DI plays a crucial role in decoupling components and promoting code reusability, maintainability, and testability.

With DI in Angular, dependencies are typically defined as services or other components that a class relies upon. These dependencies are declared in the constructor of the class, and Angular's DI system automatically resolves and injects the appropriate instances when the class is instantiated.

Here are some best practices when working with DI:

- **Singleton services**: By default, Angular services are singleton, meaning there's only one instance of the service throughout the application. This is a good practice as it ensures data consistency and optimized memory usage.
- **Use providedIn**: When providing services, use the `providedIn` property with the value `root`. This ensures that the service is available application-wide and gets tree-shaken if not used.
- **Hierarchical injectors**: Understand that Angular has a hierarchical injector system. While services provided in the root are available application-wide, services provided at a component level are available only within that component and its children.
- **Interface-based injection**: Sometimes, it's beneficial to inject a service based on an interface rather than a concrete class. This makes the system more flexible and allows for easier testing and mocking.
- **Avoid complex logic in constructors**: Since constructors are the primary place for injection, keep them clean. Avoid placing complex logic or operations that can block the component's initialization.

In the next section, we will delve into the specifics of how dependency injection works in practice.

Angular services

In Angular, a **service** is a class that provides a specific functionality or data to multiple components throughout an application. Services act as a bridge between components, facilitating the sharing of data, communication with external APIs, and performing various business logic operations. They promote code reusability, modularity, and separation of concerns within an application.

To use a service in Angular, we first need to create the service class. We can generate a new service using the `ng generate service` Angular CLI command. Once the service is created, we can inject it into any component or another service using dependency injection.

Let's take the example of a `UserService` instance that manages user-related operations, such as fetching user data from an API. Here's how we can create the `UserService` instance:

```
import { Injectable } from '@angular/core'

@Injectable({
  providedIn: 'root'
})
export class UserService {
  getUsers(): Observable<User[]> {
    // Fetch user data from API
  }
}
```

In this code, `UserService` is created as an injectable service using the `@Injectable` decorator, which indicates that this class is eligible for DI. Then the `providedIn: 'root'` option ensures that Angular creates a single instance of the service that is shared across the entire application.

Now, let's see how we can use `UserService` in a component:

```
import { Component, inject } from '@angular/core'
import { UserService } from './user.service'

@Component({
  selector: 'app-user-list',
  template: `
    <ul>
      <li *ngFor="let user of users$ | async">{{ user.name }}</li>
    </ul>
  `
})
export class UserListComponent {
  private userService = inject(UserService)
```

```
  users$ = this.userService.getUsers()
}
```

The provided code demonstrates the usage of the DI of `UserService` in an Angular component, specifically in `UserListComponent`. Let's break it down and explain its functionality:

- `private userService = inject(UserService)`: This `inject` function manually injects an instance of `UserService`, so `UserListComponent` can assess its properties and methods. You can also run `inject` inside the `constructor`.
- `users$ = this.userService.getUsers()`: This line declares the `users$` property, which takes its value from the result of the `getUser()` method from the `UserService` instance.
- `*ngFor="let user of users$ | async"`: This line enables iterating over the collection of `users$` and generating the HTML elements for each item. The `async` pipe is used in conjunction to subscribe to the `users$` observable and handle the asynchronous nature of the data. The `async` pipe automatically unsubscribes from the observable when the component is destroyed, preventing memory leaks.

> **Note**
>
> The usage of `$` in `users$` is a pattern to indicate that the property is an observable. An **observable**, a core feature from the RxJS library, is a mechanism for handling streams of asynchronous events or data over time. You can read more at `https://rxjs.dev/guide/observable`.

Angular directives

Directives are a powerful feature in Angular that allows you to extend the functionality of HTML elements. They are used to manipulate the DOM, apply custom behavior, and dynamically change the appearance or behavior of elements. Angular provides three types of directives:

- **Component directives**: Every component in Angular is a directive, such as `<app-my-component>`, which Angular interprets to create an instance of the corresponding component, encapsulating its behavior, view, and data interactions. The difference between component directives and other directives is that components contain templates.
- **Attribute directives**: These modify the behavior or appearance of an element, component, or another directive. For example, Angular's built-in `ngStyle` directive can be used to change multiple styles at the same time.
- **Structural directives**: These manipulate the layout of the DOM. For instance, Angular's built-in `*ngFor` directive can be used to render a list of items, and `*ngIf` can conditionally show or hide an element based on a boolean expression.

To use a directive in Angular, you can either leverage the built-in directives provided by Angular or create your own custom directives. Let's look at an example of using the built-in `ngIf` directive to conditionally show an element:

```
<div *ngIf="showMessage">
  <p>This message is shown conditionally.</p>
</div>
```

In the code, the `ngIf` directive is applied to the `<div>` element. The `showMessage` property is evaluated, and if it is truthy, the `<div>` element and its content will be rendered. Otherwise, they will be removed from the DOM.

You can also create your own custom directives to encapsulate specific behavior or styling by running the following command:

```
ng g d fallback-image
```

Here's an example of creating a custom directive called `FallbackImageDirective` that shows a fallback image when an existing image cannot be found:

```
import { Directive, Input, ElementRef, HostListener, inject } from '@
angular/core'

@Directive({
  selector: 'img[fallbackImage]',
  standalone: true
})
export class FallbackImageDirective {
  private el = inject(ElementRef)
  @Input() fallbackImage: string
  @HostListener('error')

  private onError() {
    const img = new Image()
    img.src = this.fallbackImage
    img.onload = () => (this.el.nativeElement.src = this.
fallbackImage)
  }
}
```

In this code, `FallbackImageDirective` is created using the `@Directive` decorator. It defines a selector, `img[fallbackImage]`, which means the directive will be applied to an image element with the `fallbackImage` attribute.

Here is an example of how to use `FallbackImageDirective`:

```
<img
  src="/does-not-exist.png"
  [fallbackImage]="'/default.png'"
/>
```

The `HostListener` decorator is used to listen to events on the host element. In this case, we listen to the `error` event and call the corresponding methods. `ElementRef` gives us access to the host element, allowing us to modify its properties, such as the `src` attribute.

Angular pipes

Pipes are a powerful feature in Angular that allows you to transform and format data right within your template. They provide a convenient way to perform data manipulation operations, such as filtering, sorting, formatting, and more. Plus, they are lightweight and reusable, and they can be chained together to create complex transformations.

Angular provides a set of built-in pipes that you can use out of the box. Let's take a look at a few examples of using these pipes:

- **The date pipe**: The `date` pipe is used to format dates. Here's an example:

  ```
  <p>Today is {{ today | date: 'longDate' }}</p>
  ```

  Here, the `today` variable represents the current date. The `date` pipe formats the date using the `'longDate'` format, which displays the date in a long format, such as `"January 1, 2023"`.

- **The currency pipe**: The `currency` pipe is used to format currency values. Here's an example:

  ```
  <p>The price is {{ price | currency: 'USD' }}</p>
  ```

  Here, the `price` variable represents a currency value. The `currency` pipe formats the value as a currency using the `'USD'` currency code, such as `"$10.99"`.

- **The uppercase pipe**: The `uppercase` pipe is used to convert a string to uppercase. Here's an example:

  ```
  <p>{{ greeting | uppercase }}</p>
  ```

In addition to the built-in pipes, you can also create your own custom pipes to perform custom data transformations. First, you can use the `ng` command to create a standalone pipe:

```
ng g p reverse
```

> **Note**
>
> If you are using an Angular version lower than v17, make sure to append `--standalone` when generating standalone pipes.

Then we can add logic to create a custom pipe called `ReversePipe` that reverses a string:

```
import { Pipe, PipeTransform } from '@angular/core';

@Pipe({
  standalone: true,
  name: 'reverse'
})
export class ReversePipe implements PipeTransform {
  transform(value: string): string {
    return value.split('').reverse().join('');
  }
}
```

In this example, a custom Angular pipe named `ReversePipe` is created using the `@Pipe` decorator. The pipe takes a string, reverses it, and returns the reversed string. The `transform` method, required by the `PipeTransform` interface, is where this reversal operation is implemented.

Then, if we want to use this custom pipe, all we have to do is import the pipe to the component and use it as normal:

```
@Component({
  selector: 'my-app',
  standalone: true,
  imports: [CommonModule, ReversePipe],
  template: `
    <p> Revere of 'abc' is {{'abc' | reverse}} </p>
  `,
})
```

In this line of code, Angular's pipe mechanism is being used to transform the string `abc` using a custom pipe named `reverse`. The line `Reverse of 'abc' is {{'abc' | reverse}}` will render as `Reverse of 'abc' is cba` in the HTML.

> **Note**
>
> Pipes are chainable, so you can combine different pipes to create the desired result. For example, `{{'abc' | reverse | uppercase}}` will reverse the string and make it uppercase. The final result is `CBA`.

Angular signals

Angular Signals is a system that granularly tracks how and where your state is used throughout an application, allowing the framework to optimize rendering updates. This feature in Angular 16 is a native way of handling reactivity in Angular.

In most cases, signals are all you need to build a simple application. A **signal** is a function that holds a value and allows you to listen to changes to that value. Let's take a look at the following example:

```
import { Component, effect, signal, WritableSignal } from '@angular/
core'

@Component({
  standalone: true,
  selector: 'my-app',
  template: `
    <p>Current count is: {{count()}}</p>
    <button (click)="setRandomCount()">Set random count</button>
  `,
})
export class App {
  public count: WritableSignal<number> = signal<number>(4);

  constructor() {
    effect(() => {
      console.log(`The current count is: ${this.count()}`);
    });
  }

  setRandomCount() {
    this.count.set(Math.floor(Math.random() * 10 + 1));
  }
}
```

This code block shows how to use signals and effects from the `@angular/core` package to manage state and side effects. Here's a breakdown:

- `public count: WritableSignal<number> = signal<number>(4)`: This line declares a signal `count`, which is a `WritableSignal` instance that holds a number. The initial value of `count` is set to `4`.
- `setRandomCount() { ... }`: This is a method that sets `count` to a random number between 1 and 10. The `this.count.set(...)` method is used to update the value of `count`.
- `effect()` is an operation that runs whenever one or more signal values change and prints the current value of `count` to the browser console log.

- The template of the component displays the current value of `count` and a button that calls `setRandomCount` when clicked.

> **Note**
>
> Angular Signals is marked as stable since Angular 17. You can read more on Angular Signals at `https://angular.io/guide/signals`.

Angular control flow

In an effort to enhance the **developer experience** (**DX**), Angular 17 introduced a new feature called **Built-in control flow**. This feature allows you to use a familiar JavaScript-like syntax in Angular templates to easily show, hide, or repeat elements.

Let's take the `NgIf` directive as an example. Suppose we have a list of products and we want to display different content when no products are available. Before the introduction of the control flow feature, here's how we would handle it:

```
<div *ngIf="products.length; else noProducts">
  Show product list
</div>
<ng-template #noProducts>
  Products are empty!
</ng-template>
```

Overall, this code snippet displays a `Show product list` message inside a `<div>` element if the `products` array has elements. If the `products` array is empty, the `Products are empty!` message from the `ng-template` block is displayed instead. This allows for conditional rendering based on the state of the `products` array.

Now let's see how it's done with the new `@if` control flow:

```
@if (products.length) {
  Show Product List
} @else {
  Products are empty!
}
```

In this code snippet, you'll notice that we can directly utilize `@if`, `@else`, and even `@else if` syntax, resulting in cleaner and more familiar code. This approach also exhibits a JavaScript-like coding style, enhancing the DX with a sense of familiarity and ease.

> **Note**
>
> In addition to enhancing the DX, the new control flow also significantly improves the performance of your application compared to the previous implementation. You can refer to the community framework benchmarks available at `https://krausest.github.io/js-framework-benchmark/current.html` for further verification.

The Angular team also provides a nice way to run migration to move existing syntax to the new control flow. All you need to do is to run the following command:

```
ng generate @angular/core:control-flow
```

This command will scan the existing implementations of `NgIf`, `NgFor`, and `NgSwitch` and upgrade them to the new control flow. It is important to thoroughly test and validate that everything functions as intended before finalizing and committing these changes.

> **Note**
>
> Angular control flow is still under Developer Preview in Angular 17. You can read more on Angular control flow at `https://angular.io/guide/control_flow`.

Up to this point, we have explored several core features of Angular that serve as fundamental building blocks for developing functional Angular applications. In the next section, we will delve into valuable tips on effectively organizing your Angular project.

Organizing an Angular project

The purpose of organizing and structuring an Angular application is to enhance its maintainability, scalability, and reusability. It involves making decisions about how to structure the code base, divide responsibilities among different files and folders, and establish conventions for naming and organizing files. Organizing an Angular project effectively not only improves the developer experience but also helps teams collaborate better and reduces the learning curve for new developers joining the project.

When organizing an Angular project, it's important to follow established best practices and leverage Angular's recommended project structure. The Angular style guide provides several best practices for organizing an Angular project. You can find the style guide here: `https://angular.io/guide/styleguide`.

One key principle emphasized in the style guide is the **LIFT approach**, which stands for *Locating* code quickly, *Identifying* the code at a glance, *Flattening* the structure, and *Trying* to be **DRY** (**Don't Repeat Yourself**). Let's go through each aspect of the LIFT principle and provide a code example for better understanding:

- **Locating code quickly**: The goal is to organize the code base in a way that allows you to locate files and modules easily. One common approach is to group files based on their functionality or features. Here's an example:

```
app/
  components/
    product/
      product.component.ts
      product.component.html
      product.component.scss
      product.component.spec.ts
  services/
    product.service.ts
  models/
    product.model.ts
```

  Here, related files are grouped together within directories such as `components`, `services`, and `models`. This structure helps developers quickly find the files they need when working on a specific feature.

- **Identifying the code at a glance**: The focus is on using meaningful and descriptive names for files, classes, variables, and functions. This makes it easier to understand the purpose and functionality of the code. Here's an example:

```
// product.component.ts
export class ProductComponent {
  // Component logic...
}
```

  Using descriptive names such as `ProductComponent` clearly communicates the responsibility and purpose of the code.

- **Flattening the structure**: The aim is to keep a flat directory structure instead of nesting directories too deeply. This simplifies navigation and maintenance. Here's an example:

```
app/
  components/
    product/
      product.component.ts
```

  Here, the `product.component.ts` file is located directly under the `components` directory without unnecessary nesting.

- **Trying to be DRY**: The principle encourages code reuse and avoiding duplication. It promotes extracting common functionality into reusable components, services, or modules. Here's an example:

```
// shared/ui/loading-spinner.component.ts
```

```
@Component({
  selector: 'app-loading-spinner',
  template: `
    <div class="loading-spinner">
      <!-- Loading spinner template -->
    </div>
  `,
})
export class LoadingSpinnerComponent {
  // Component logic...
}
```

`LoadingSpinnerComponent` can be reused in multiple parts of the application to display a loading spinner, reducing code duplication, and improving consistency.

By adhering to the LIFT principle, Angular projects can benefit from a well-organized and maintainable code base, making development more efficient and enhancing collaboration among team members.

In addition to the LIFT approach, the style guide also provides the following best practices which are worth noting here:

- **Follow the Angular CLI workspace structure**: The Angular CLI provides a recommended project structure that aligns with the style guide. It separates source files from configuration files and organizes them under folders such as `src/app` for application-specific code and `src/assets` for assets.
- **Separate concerns with directories**: Use directories to separate concerns and group related files. For example, place components, templates, styles, and tests for a feature in the same directory. This makes it easier to locate and maintain related code.
- **Name files consistently**: Use consistent file naming conventions. For example, use `feature-name.component.ts` for component files, `feature-name.service.ts` for service files, and `feature-name.spec.ts` for test files. This consistency helps developers locate and identify files quickly.
- **Use barrels**: Create barrel files (`index.ts`) in directories to provide a central entry point for exporting multiple files. **Barrels** simplify imports and make it easier to locate and manage related files.
- **Separate root and feature-level code**: Keep root-level code separate from feature-level code. Place code that is used throughout the application, such as the root component, routing, and global services, in the root module (`AppModule`); feature-level code should be placed in feature modules.
- **Consider a shared directory**: Create a shared directory to encapsulate commonly used components, directives, pipes, and services. The shared module simplifies importing shared resources and promotes code reuse.

- **Use lazy loading**: For large applications, consider using **lazy loading** to load feature modules on demand. This improves initial load times and separates code into manageable chunks.

Remember that the Angular style guide provides detailed explanations and examples for each of these best practices, and it's a valuable resource for organizing an Angular project effectively.

> **Note**
>
> These are just recommendations on best practices when organizing an Angular project. There's no *one size fits all* approach in terms of architecting your Angular project. It will always depend on other factors such as the stage of the project, deadline, team size, and so on. Make sure that you pick the approach that is suitable for your project. A recommended strategy is to begin with the LIFT approach, which facilitates faster application development. As your application expands, it is advisable to apply best practices to optimize your code and ensure its efficiency.

Summary

In this chapter, we have covered key aspects of modern Angular development. We began by introducing the evolution of Angular, highlighting its growth and advancements. Moving on, we discussed core features and improvements in recent Angular versions, including data binding, components, services, directives, pipes, and signals. We also emphasized the importance of organizing an Angular project effectively using best practices.

By gaining insights into modern Angular development, you are now equipped to take your skills to the next level. In the next chapter, we will focus on integrating PrimeNG into your Angular project. We will guide you through the process of incorporating PrimeNG into your Angular application and harnessing its power to enhance the user experience.

4

Integrating PrimeNG into Your Angular Project

In this chapter, we will explore the process of integrating PrimeNG, a popular UI component library, into your Angular projects. PrimeNG provides a rich set of pre-built components that can enhance the functionality and aesthetics of your application. Whether you need to incorporate complex data tables, responsive layouts, or interactive charts, PrimeNG offers a wide range of components to meet your requirements. By utilizing PrimeNG, it will help to save time from building foundation components in a robust, reliable, and accessible way, so that you can focus on the application rather than building all the core components of typical UI elements.

By following the step-by-step instructions and examples provided in this chapter, you will gain the knowledge and skills necessary to seamlessly integrate PrimeNG into your Angular projects, and create feature-rich and visually appealing applications with ease.

The chapter will cover the following topics:

- Adding PrimeNG components to your Angular project
- Configuring PrimeNG modules and dependencies
- Working with PrimeNG component APIs and directives
- Customizing component styles and themes
- Troubleshooting common integration issues

Technical requirements

This chapter contains various working code samples on how to integrate PrimeNG into Angular projects. You can find the related source code in the `chapter-04` folder of the following GitHub repository: `https://github.com/PacktPublishing/Next-Level-UI-Development-with-PrimeNG/tree/main/apps/chapter-04`.

Adding PrimeNG components to your Angular project

PrimeNG is a powerful UI component library for Angular that provides a wide range of pre-built components to enhance the functionality and visual appeal of your application. In this section, we will explore how to add PrimeNG and its dependencies to your Angular project, import the necessary styles, and utilize PrimeNG components in your templates.

Installing PrimeNG

Before starting to integrate PrimeNG into your Angular project, make sure that you set up a fresh project to begin with. Now follow these steps to install PrimeNG in your Angular project:

1. Open a terminal or command prompt in your project directory.
2. Run the following command to install PrimeNG and save it as a dependency in your project:

   ```
   npm install primeng
   ```

After the installation process, you will see `primeng` in the `package.json` file in your root directory.

> **Note**
>
> At the time of writing this book, we're using PrimeNG version `17.0.0`. If you have a different version and your app doesn't work correctly, you can try with `yarn add primeng@17.0.0` to install the correct version of the plugin for this book.

The `primeng` package is all that you need to get started with integration. In the next section, we will add some styles and themes to your application.

Importing PrimeNG styles into your Angular application

The Theme and Core styles are essential CSS files for the components. You can find the comprehensive selection of available themes in the **Theming** section at `https://primeng.org/theming#themes`. To incorporate the styles, you can import them either in the `angular.json` or `src/styles.css` file. In this section, we will use the `lara-light-blue` theme as an example.

In order to add themes and styles to your `angular.json` file, go to the `styles` section of the file and add the styles like so:

```
// angular.json

"options": {
    ...
    "styles": [
        "apps/chapter-04/src/styles.scss",
```

```
        "node_modules/primeng/resources/themes/lara-light-blue/theme.
css",
        "node_modules/primeng/resources/primeng.min.css"
    ],
},
```

Besides adding themes and styles to the `angular.json` file, you can also utilize `styles.scss` to import the styling:

```
// styles.scss

/* You can add global styles to this file, and also import other style
files */
@import 'primeng/resources/themes/lara-light-blue/theme.css';
@import 'primeng/resources/primeng.css';
```

In addition, each theme has its own font family; it's suggested to apply a font family to your application to achieve a unified look:

```
// styles.scss

body {
        font-family: var(--font-family);
}
```

> **Note**
>
> `--font-family` is a CSS variable, also known as a CSS custom property, which are placeholders that hold values and can be used throughout a CSS stylesheet. You can learn more at `https://developer.mozilla.org/en-US/docs/Web/CSS/Using_CSS_custom_properties`.

As an example, for the `lara-light-blue` theme, the default font family is defined under the `theme.css` file:

```
// node_modules/primeng/resources/themes/lara-light-blue/theme.css

:root {
    ...
    --font-family:-apple-system, BlinkMacSystemFont, Segoe UI, Roboto,
Helvetica, Arial, sans-serif, Apple Color Emoji, Segoe UI Emoji, Segoe
UI Symbol;
    ...
}
```

> **Note**
>
> If you are using Nx Workspace, instead of `angular.json`, use `project.json`.

Working with PrimeNG icons

Icons play a crucial role in enhancing the visual appeal and usability of web applications, providing intuitive visual cues and helping users quickly identify and interact with various elements. In this section, we will explore how to work with icons in PrimeNG and leverage the vast collection of icons it offers.

To utilize PrimeIcons, you need to install the `primeicons` package:

```
npm install primeicons
```

After that, the CSS file of the icon library needs to be imported in `styles.scss` of your application:

```
// style.scss
@import "primeicons/primeicons.css";
```

> **Note**
>
> For a full list of PrimeIcons, please check this website: `https://primeng.org/icons`.

PrimeIcons offers special syntax: `pi pi-{icons}`, which can be utilized in your Angular components. You have the flexibility to substitute `{icons}` with an icon name from the provided link, such as `pi pi-user`, to indicate that it represents a user icon. In case you want to use Prime Icons as a standalone element, you can utilize the `i` or `span` element, like so:

```
<i class="pi pi-user"></i>
<span class="pi pi-user"></span>
```

In this example, we use the `pi-user` icon. You can replace it with any other icon name from the supported icon libraries.

Adding PrimeNG methods

There are two methods to add PrimeNG to your template: PrimeNG components and PrimeNG directives. To illustrate this, let's use PrimeNG's `p-button` element.

The `p-button` component is a self-contained UI element that can be customized and reused throughout your app. It comes with built-in functionality, such as event handling, and a variety of customization options, including icon support, label, and style classes. Here's an example of how to use the `p-button` component:

```
<p-button label="Click me" />
```

In contrast, the pButton directive is used to add behavior to an existing HTML button element. It allows you to enhance the functionality of the button element by adding PrimeNG-specific features, such as styling and event handling. Here's an example of how to use the pButton directive:

```
<button pButton type="button" label="Click me"></button>
```

In this example, we're using the pButton directive to enhance an existing HTML button element with PrimeNG-specific features. We're setting the type attribute to "button" and adding a label to the button.

It's recommended to use the p-button component when you need a self-contained UI element that can be customized and reused throughout your app, and when you need to add a lot of built-in functionality to your button. For the pButton directive, you can use it to add behavior to existing HTML button elements such as button, input, or tag, and when you need to enhance the functionality of the button with PrimeNG-specific features.

Using PrimeNG components in your templates

The section focuses on incorporating PrimeNG components seamlessly into your templates. We will learn how to leverage the extensive library of PrimeNG components effectively within your web applications. Let's get started.

If you are using the NgModule approach, all you need to do is import the PrimeNG modules to NgModule decorator:

```
import { NgModule } from '@angular/core'
import { ButtonModule } from 'primeng/button'

@NgModule({
   imports: [
      // Other module imports
      ButtonModule
   ],
   // Other module configurations
})
export class AppModule { }
```

The previous code shows that we imported ButtonModule into AppModule.

Once the PrimeNG modules are imported, you can start using PrimeNG components in your Angular templates. Follow these steps to utilize PrimeNG components:

1. Open a component template file (.html) where you want to use PrimeNG components.

2. Place the PrimeNG component selector in the template where you want the component to appear. For example, to add a button component, use the following code:

```
<p-button label="Click me" />
```

For the standalone component approach, you will import the PrimeNG component into the Angular component that you are working on:

```
import { Component } from '@angular/core'
import { ButtonModule } from 'primeng/button'

@Component({
   standalone: true,
   imports: [ButtonModule],
   selector: 'primengbook-root',
   template: `
      <h1>Welcome to chapter-04</h1>
      <p-button label="Click me" />
   `
})
export class AppComponent {}
```

In the previous code snippet, we have `AppComponent`, which utilizes `ButtonModule` from PrimeNG. `p-button` is an indicator that it's a PrimeNG button component.

As a result, you will have a styled button on your browser:

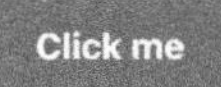

Figure 4.1 – PrimeNg button component

Working with PrimeNG component APIs and directives

As a developer, you understand the importance of leveraging powerful and customizable libraries to streamline your development process. Here, we'll see how to use PrimeNG APIs and directives to interact with components programmatically, customize their behavior, and harness their full potential.

Each PrimeNG component will have its own APIs and configurations. Let's consider the PrimeNG button component as an example:

```
<p-button
  label="Click me"
  icon="pi pi-check"
  iconPos="right"
```

```
    [disabled]="isDisabled"
/>
```

In this example, we're using the `label` input to set the button label, the `icon` input to set the icon displayed on the button, the `iconPos` input to set the icon position to the right of the label, and the `disabled` input to disable the button based on the value of `isDisabled` property.

You can see the result in the following screenshot:

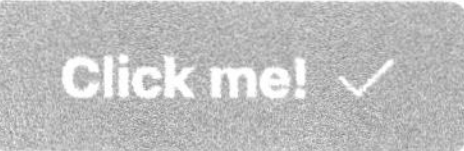

Figure 4.2 – PrimeNG button configuration

Here are some other configuration options you can use with `p-button`:

- `styleClass`: Sets the CSS class for the button
- `type`: Sets the type of button (e.g. `"button"`, `"submit"`, `"reset"`)
- `tabIndex`: Sets the tab index for the button
- `loading`: Indicates the button is in the loading state
- `loadingIcon`: Sets the icon to display in the loading state

In case you want to trigger an event after clicking the button, you can utilize the `click` event emitter in order to trigger a function from your component:

```
<button
   pButton
   label="Click me!"
   [loading]="loading"
   (click)="onClickEvent()"
></button>

...

onClickEvent() {
   this.loading = true

   setTimeout(() => {
      this.loading = false
   }, 2000)
}
```

The following code shows a PrimeNG button with a click event handler:

- `(click)="onClickEvent()"`: When the user clicks on the button, it will trigger the `onClickEvent()` function on the component
- `[loading]="loading"`: Show the loading state of the button after being clicked

Let's take a look at the result:

Figure 4.3 – PrimeNG button with click event

After the button is clicked, the loading state is activated and shows a loading icon on the left of the button.

By following these steps, you can easily add PrimeNG components to your Angular project. In the next section, we will delve into some PrimeNG configurations.

Configuring PrimeNG modules and dependencies

PrimeNG provides a comprehensive set of configuration options that can be applied to its components. These options enable you to control various aspects of the component's functionality and visual presentation. Let's take a look at some commonly used configuration options.

Global configuration

PrimeNG offers a global configuration object that allows you to define default settings for all components in your application. For example, you can configure the default locale, animation duration, and other global options using the `PrimeNGConfig` object. The provided code snippets show the initialization and configuration of PrimeNG during the bootstrapping process.

First, Let's create the global configuration file for PrimeNG:

```
// shared/providers/primeng.provider.ts

import { APP_INITIALIZER } from '@angular/core'
import { PrimeNGConfig } from 'primeng/api'

const factory = (primengConfig: PrimeNGConfig) => {
  return () => {
    primengConfig.ripple = true
```

```
        primengConfig.zIndex = {
            modal: 1100,        // dialog, sidebar
            overlay: 1000,     // dropdown, overlay panel
            menu: 1000,         // overlay menus
            tooltip: 1100,     // tooltip
        }

        // more configuration options
    }
}

export const primeNgProvider = {
    provide: APP_INITIALIZER,
    useFactory: factory,
    deps: [PrimeNGConfig],
    multi: true,
}
```

This code defines the `primeNgProvider`, which is a provider that initializes and configures PrimeNG. It uses the `APP_INITIALIZER` token from `@angular/core` to ensure that the configuration is applied before the application starts.

> **Note**
>
> `APP_INITIALIZER` is an Angular feature that allows you to run some initialization tasks before your application starts. By providing an array of functions or services to the `APP_INITIALIZER` token in the `providers` array of your application configuration, Angular ensures that these tasks are executed and completed before the application is fully loaded. You can learn more at `https://angular.io/api/core/APP_INITIALIZER`.

The `factory` function is the implementation of the provider. It receives the `PrimeNGConfig` object as a dependency and returns a function. When this function is executed during the application initialization, it configures various aspects of PrimeNG. In this example, it enables the ripple effect (`ripple: true`), which is a visual animation or graphical effect that occurs when the button is interacted with and sets the `z-index` values for different PrimeNG components.

After that, we can add the PrimeNG global config to the app configuration file:

```
// app.config.ts

import { ApplicationConfig } from '@angular/core'
import { primeNgProvider } from './shared/providers'

export const appConfig: ApplicationConfig = {
  providers: [primeNgProvider],
}
```

This code is part of the Angular application's bootstrap process. It provides the `primeNgProvider` as a provider, which is responsible for initializing and configuring PrimeNG.

By providing the `primeNgProvider` as a provider in the application's bootstrap process, it ensures that PrimeNG is properly initialized and configured before the application starts, allowing components and features of PrimeNG to be used seamlessly throughout the application.

Angular animation configuration

Animations play a crucial role in creating engaging and interactive user interfaces. In Angular, animations can be utilized to bring components to life and provide a smooth and visually appealing user experience. In this section, we will explore how to integrate and configure animations in your Angular application.

To enable animations in your Angular application, you need to import the `provideAnimations` function from the `@angular/platform-browser/animations` package. This function provides the necessary functionality and tools to work with animations in Angular. It leverages the underlying animation capabilities of the browser to deliver seamless animations.

To import `provideAnimations`, you need to include it in the `providers` array of your app configuration:

```
// app.config.ts

import { ApplicationConfig } from '@angular/core'
import { provideAnimations } from '@angular/platform-browser/
animations'

export const appConfig: ApplicationConfig = {
  providers: [
    ...
    provideAnimations()
  ],
}
```

In certain scenarios, you may prefer to disable animations globally in your application. This could be to optimize performance or accommodate specific user preferences. Angular provides the `provideNoopAnimations` function as an alternative to `provideAnimations` for disabling animations:

```
// app.config.ts

import { ApplicationConfig } from '@angular/core'
import { provideNoopAnimations } from '@angular/platform-browser/
animations'

export const appConfig: ApplicationConfig = {
   providers: [
      ...
      provideNoopAnimations()
   ],
}
```

The `provideNoopAnimations` function essentially replaces the animation functionality with a no-operation implementation. This results in animations being disabled throughout your application.

In the next section, we will explore further how to customize styles and themes.

Customizing component styles and themes

PrimeNg provides a wide range of components with default styles that are designed to be functional and visually appealing. However, to create a seamless integration with your application's design, you may need to customize the appearance of PrimeNG components. In this section, we will explore various techniques for customizing component styles and themes in PrimeNG to achieve the desired look and feel.

Overriding styles at the component level

Let's say you want to change the background color and border radius of a `Button` component. Here's an example of how you can override the default styles of the `Button` component using CSS:

```
import { CommonModule } from '@angular/common'
import { Component } from '@angular/core'
import { ButtonModule } from 'primeng/button'

@Component({
   selector: 'primengbook-button-override-styles',
   standalone: true,
   imports: [CommonModule, ButtonModule],
```

```
    template: `
      <h2>Button Override Styles</h2>

      <button pButton type="button" label="Custom styles"></button>
    `,
    styles:
      `
        .p-button {
          background-color: #696cff;
          border-radius: 5px;
        }
      `,
})
export class ButtonOverrideStylesComponent {}
```

In this example, we're targeting the `.p-button` class, which is the class applied to all `Button` components. We're then using the `background-color` and `border-radius` properties to change the appearance of the `Button` component. Let's take a look at the result:

Figure 4.4 – Component custom button styles

If you inspect the browser, you will see that the styling is being applied correctly:

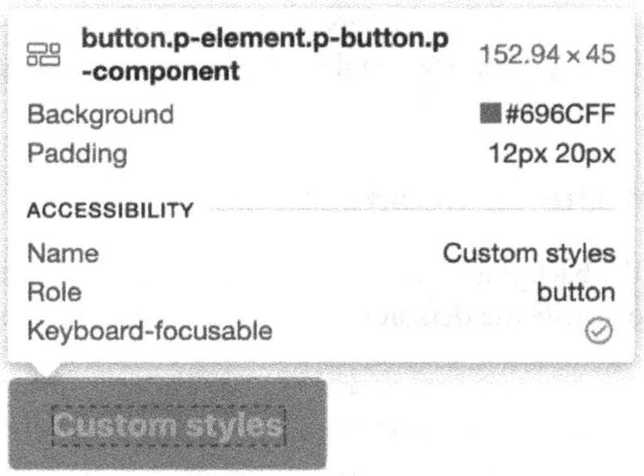

Figure 4.5 – Component custom button style inspection

You can see from the inspection the background color of the button has the color code `#696CFF`, which is the one that we set in the styles section.

Overriding styles globally

In the previous section, we learned how to add custom styles at a component level. In this part, we will learn how to apply styling globally using **CSS preprocessors**. Before we begin, let's quickly look into what CSS preprocessors are.

CSS preprocessors are scripting languages that extend the default capabilities of **Cascading Style Sheets** (**CSS**). They enable developers to use variables, nested rules, mixins, functions, and more, which can make CSS more maintainable, themeable, and extendable:

- **Variables**: These allow you to store values that you want to reuse throughout your style sheet. For example, you can define a variable for your primary color and use it in multiple places:

  ```
  $primary-color: #5d738a;

  .p-button {
     background-color: $primary-color;
  }
  ```

  This styling will apply all the new background colors to all PrimeNG buttons.

- **Nested rules**: These let you nest CSS selectors inside other selectors, making your CSS more readable and maintainable:

  ```
  nav {
     ul {
        margin: 0;
        padding: 0;
        list-style: none;
     }

     li {
        display: inline;
        margin-right: 5px;
     }
  }
  ```

- **Mixins and functions**: These enable you to define reusable chunks of CSS. They can take parameters, allowing you to customize the output:

  ```
  @mixin box-shadow($x, $y, $blur, $color) {
        -webkit-box-shadow: $x $y $blur $color;
            -moz-box-shadow: $x $y $blur $color;
                   box-shadow: $x $y $blur $color;
  }
  ```

```
button {
      @include box-shadow(0, 0, 5px, #ccc);
}
```

Inside the mixin, there are three declarations, each representing a vendor-prefixed version of the box-shadow property. The vendor prefixes are *-webkit-* for WebKit-based browsers (e.g., Chrome and Safari), and *-moz-* for Mozilla-based browsers (e.g., Firefox). The final declaration is the standard `box-shadow` property.

Wherever you want to apply a box shadow to your element, you can just include the `box-shadow` mixin. It will generate the following code for you:

```
button {
      -webkit-box-shadow: 0 0 5px #ccc;
          -moz-box-shadow: 0 0 5px #ccc;
                  box-shadow: 0 0 5px #ccc;
}
```

> **Note**
>
> Using a CSS preprocessor can greatly enhance your workflow, especially in large projects. However, there's a learning curve, and you must set up your development environment to compile the code.

Now, let's take a look at an example of how you can apply global styling:

```
// styles.scss

@import 'primeng/resources/themes/lara-light-blue/theme.css';
@import 'primeng/resources/primeng.css';

@import 'primeicons/primeicons.css';


:root {
   --font-family: ui-serif, Georgia, Cambria, 'Times New Roman',
Times, serif;
}

// Define a variable for primary color
$primary-color: #5d738a;
$text-color: #f8f4ef;

// Define box shadow
@mixin box-shadow($x, $y, $blur, $color) {
```

```
        -webkit-box-shadow: $x $y $blur $color;
            -moz-box-shadow: $x $y $blur $color;
                    box-shadow: $x $y $blur $color;
}

.p-button {
    background-color: $primary-color;
    color: $text-color;
    @include box-shadow(0, 0, 5px, #ccc);
}
```

In this example, we're using a CSS variable to override `font-family` from `lara-light-blue/theme.css`. After that, we add a Sass variable to define the primary color and text color values. Then we add the `box-shadow` property via the mixin.

We're using these variables to style the `Button` component globally. Let's check the changes:

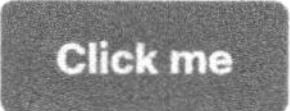

Figure 4.6 – Global custom button styles

After debugging the browser, you can see that except for the modified styling button at the component level, all other buttons in your applications are updated with new styling globally.

> **Note**
>
> When customizing component styles or applying themes, be sure to test your changes thoroughly to ensure that they don't adversely affect the functionality or usability of your application. Also, be aware that some components may have specific CSS classes or styles that you need to target in order to customize their appearance.

Customizing the appearance of PrimeNG components is an important part of creating a visually appealing and cohesive web application. Whether you're using CSS, CSS preprocessors, or PrimeNG's built-in themes, there are a range of tools and customization options available to help you achieve your design goals. By taking the time to customize your application's styles and themes, you can create an application that is both functional and aesthetically pleasing.

In the next section, we will explore common integration issues that may arise when working with PrimeNG components and provide troubleshooting tips to overcome them.

Troubleshooting common integration issues

Integrating PrimeNG into your Angular project can sometimes come with its own set of challenges. In this section, we will explore common issues that may arise during PrimeNG integration and provide troubleshooting techniques to help you overcome them. By understanding these challenges and having the necessary debugging skills, you can ensure a smooth integration process.

Conflict or compatibility issues

One common issue when integrating PrimeNG is encountering conflicts or compatibility issues with other libraries or Angular versions. It's essential to ensure that all dependencies, including Angular and PrimeNG, are compatible with each other. For instance, if you are using Angular 15.x and have installed PrimeNG 16.x, there may be conflicts when integrating PrimeNG components.

> **Note**
>
> The following is release cycle information from the PrimeNG official website: "*PrimeNG release cycle is aligned with Angular and every 6 months a new major PrimeNG version is released as open source that is compatible with the latest Angular core*". You can check the current **Long Term Support** (**LTS**) details at `https://primeng.org/lts`.

In case of conflicts, you may need to investigate the specific conflicting versions and find workarounds or update the versions accordingly. You should pay attention to the `package.json` file to check the version of your dependencies:

```
// package.json

...
"dependencies": {
      "@angular/animations": "17.0.6",
      "@angular/cdk": "17.0.3",
      "@angular/common": "17.0.6",
      "@angular/compiler": "17.0.6",
      "@angular/core": "17.0.6",
      "@angular/forms": "17.0.6",
      "@angular/platform-browser": "17.0.6",
      "@angular/platform-browser-dynamic": "17.0.6",
      "@angular/router": "17.0.6",
      "install": "^0.13.0",
      "primeflex": "^3.3.1",
      "primeicons": "^6.0.1",
      "primeng": "17.0.0",
      "rxjs": "~7.8.0",
```

```
        "tslib": "^2.3.0",
        "zone.js": "0.14.2"
},
...
```

As you can see, primeng and angular are using the same version 17.x. Having different versions doesn't mean that there will be conflicts in your Angular applications. In case you do get errors, please check versions from the official Angular and PrimeNG websites: https://github.com/angular/angular/releases and https://github.com/primefaces/primeng/releases.

After you check the release version of Angular or PrimeNG that you want to install, use npm to update it:

```
npm install primeng@17.0.0
```

This command will update primeng to version 17.0.0.

Missing or incorrect imports

Another frequent issue is missing or incorrect imports of PrimeNG modules or components. When using PrimeNG components in your templates or code, it's crucial to import the necessary modules correctly. If you forget to import a required module, the component may not work as expected or throw errors. Double-check your imports and make sure all required PrimeNG modules are imported into your Angular application.

Let's take a look at the following component:

```
@Component({
    selector: 'primengbook-button-override-styles',
    standalone: true,
    imports: [CommonModule],
    template: `
        <h2>Button Override Styles</h2>

        <button pButton type="button" label="Custom styles"></button>
    `
})
export class ButtonOverrideStylesComponent {}
```

In the previous code, you can see that we're using the pButton directive approach, and there is no error from the VS Code or the compiler. However, the button will not show correctly since you haven't imported ButtonModule from primeng into the imports array. Here is how you fix this issue:

```
import { ButtonModule } from 'primeng/button'

...
```

```
@Component({
    selector: 'primengbook-button-override-styles',
    standalone: true,
    imports: [CommonModule, ButtonModule],
```

Now the PrimeNG `ButtonModule` dependency is correctly added to the `imports` array in the component.

Incorrect configuration or setup

Sometimes, issues arise due to incorrect configuration or setup of PrimeNG features. For example, if you're using PrimeNG's animation features, ensure that you have added the `provideAnimations` or `provideNoopAnimations` function as required.

Additionally, verify that any necessary configuration options are set correctly.

Do refer to the PrimeNG documentation for detailed instructions on setting up and configuring specific features: `https://primeng.org/installation`

Inspecting console errors and warnings

When facing integration issues, the browser's developer console is an invaluable tool for debugging. It provides error messages, warnings, and additional information that can help you identify the root cause of the problem. Inspect the console for any error messages related to PrimeNG components or modules. These error messages often provide valuable insights into the issue at hand. Let's have a look at the following error:

```
▶ [webpack-dev-server] ERROR                                    index.js:493
apps/chapter-04/src/app/app.component.ts:14:5 - error NG8001: 'p-
button' is not a known element:
1. If 'p-button' is an Angular component, then verify that it is
included in the '@Component.imports' of this component.
2. If 'p-button' is a Web Component then add
'CUSTOM_ELEMENTS_SCHEMA' to the '@Component.schemas' of this
component to suppress this message.

14      <p-button label="Click me" />
```

Figure 4.7 – Sample error in the console log

Based on the error displayed in the console, the issue appears to be on line 14 in the `app.component.ts` file. The error message indicates that `p-button` is not an Angular component and suggests that you add it to the `imports` array or include `CUSTOM_ELEMENTS_SCHEMA` in the `schemas` array. To resolve this issue, you should add the PrimeNG `ButtonModule` dependency to the component's `imports` array.

Using the Angular CLI

The Angular CLI offers various helpful commands that can aid in troubleshooting. This includes using the `ng build` command to check for build errors or the `ng serve` command to run your application and observe any runtime issues.

The CLI also provides options for generating component and module schematics, which can assist in setting up PrimeNG components correctly:

```
Error: apps/chapter-04/src/app/components/button-configuration.
component.ts:11:5 - error NG8001: 'p-button' is not a known element:
1. If 'p-button' is an Angular component, then verify that it is
included in the '@Component.imports' of this component.
2. If 'p-button' is a Web Component then add 'CUSTOM_ELEMENTS_SCHEMA'
to the '@Component.schemas' of this component to suppress this
message.

11        <p-button
             ~~~~~~~~~
12           label="Click me!"
     ~~~~~~~~~~~~~~~~~~~~~~~~
...
15           [disabled]="isDisabled"
     ~~~~~~~~~~~~~~~~~~~~~~~~~~~~~~
16        />
     ~~~~~~
```

The previous output from the Angular CLI indicates that we forgot to import `ButtonModule` before using `p-button`.

Seeking help from the community

In case you encounter an issue that seems difficult to resolve, don't hesitate to seek help from the developer community. Online forums, discussion boards, and social media groups dedicated to Angular and PrimeNG are excellent resources for getting assistance. Many experienced developers are willing to share their insights and provide guidance on troubleshooting specific integration issues.

You can find the dedicated PrimeNG discussions at `https://github.com/orgs/primefaces/discussions`.

Summary

In this chapter, we explored the process of integrating PrimeNG into an Angular project. We learned how to add PrimeNG components to our application, configure PrimeNG modules and dependencies, work with PrimeNG component APIs and directives, customize component styles and themes, and troubleshoot common integration issues. By successfully integrating PrimeNG, we can leverage its rich set of UI components and features to enhance our Angular applications.

Through the chapter, we gained valuable knowledge and skills that are essential for professional developers. Integrating PrimeNG into an Angular project opens up a world of possibilities for creating feature-rich and visually appealing web applications. By harnessing the power of PrimeNG, we can save development time and effort by utilizing pre-built, customizable components, and tapping into advanced functionalities.

In the next chapter, we will delve into the realm of input components and form controls provided by PrimeNG. We will explore how to work with text inputs, checkboxes, radio buttons, dropdowns, and more, enabling us to create interactive and user-friendly forms. Additionally, we will dive into form validation techniques and learn how to handle user input effectively.

Part 2: UI Components and Features

In this part, you will dive deep into the world of PrimeNG's UI components and explore the rich set of features they offer. You will learn how to leverage these components to build dynamic and interactive user interfaces for your Angular applications.

By the end of this part, you will have a comprehensive understanding of most of PrimeNG's UI components and be able to effectively utilize them to enhance your application's functionality and user experience.

This part contains the following chapters:

5

Introducing Input Components and Form Controls

Diving deeper into the world of Angular and PrimeNG, we're about to embark on a journey through the realm of input components and form controls.

This chapter is dedicated to providing a comprehensive understanding of how to effectively utilize various input components and form controls in your Angular applications. We'll be exploring the use of text inputs, checkboxes, radio buttons, dropdowns, and more. Additionally, we'll delve into the intricacies of form validation and handling user input.

By harnessing the power of these input components and mastering form controls, we will be able to create intuitive and user-centric applications, which is paramount in today's digital landscape.

The chapter will cover the following topics:

- Introducing input components and form controls
- Working with text inputs, checkboxes, and radio buttons
- Using dropdowns, multi-selects, and date pickers
- Implementing form validation

Technical requirements

This chapter contains various working code samples on input components and Angular forms. You can find the related source code in the `chapter-05` folder of the following GitHub repository: `https://github.com/PacktPublishing/Next-Level-UI-Development-with-PrimeNG/tree/main/apps/chapter-05`.

Introducing input components and form controls

Before getting into the main content of this chapter, let's set the stage with an overview of input components and form controls. Angular provides two ways to handle user inputs through forms: template-driven and reactive forms. Both methods have their unique strengths, and choosing between them depends on the specific needs of your application.

Let's have a look at a simple Angular form:

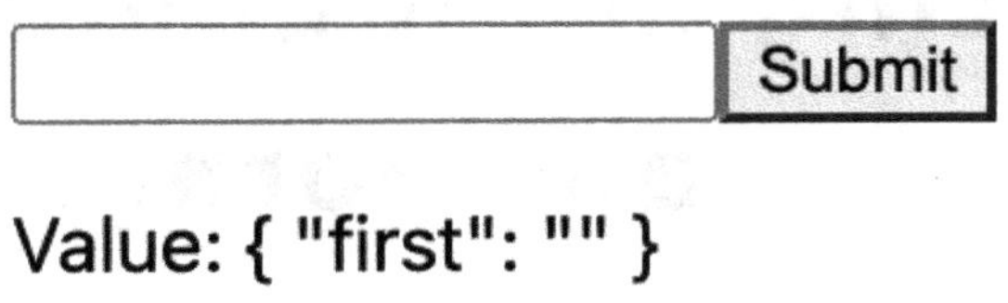

Figure 5.1 – Angular sample form

We will use both approaches to recreate this form.

Template-driven forms

Template-driven forms define controls directly within the DOM and then link them back to the underlying model. They shine in their simplicity, making them a go-to choice for straightforward use cases. For instance, when dealing with a form that has a small number of fields and uncomplicated validation rules, template-driven forms offer a rapid and effortless approach to implementing form functionality.

Using template-driven forms, here is how we would recreate *Figure 5.1*:

```
import { CommonModule } from '@angular/common'
import { Component } from '@angular/core'
import { FormsModule, NgForm } from '@angular/forms'

@Component({
  standalone: true,
  imports: [CommonModule, FormsModule],
  template: `
    <h2>Template Driven Form</h2>

    <form #form="ngForm" (ngSubmit)="onSubmit(form)" novalidate>
      <input name="first" ngModel required />
      <button>Submit</button>
    </form>
```

```
    <p>Value: {{ form.value | json }}</p>
  `
})
export default class TemplateDrivenComponent {
  onSubmit(form: NgForm) {
    console.log(form.value) // { first: '' }
  }
}
```

Here's a breakdown of important parts of the code:

- `FormsModule`: This module is imported when you want to use template-driven forms in your Angular application. It provides directives such as `NgForm` and `NgModel`:
  - `NgForm`: This is a directive that automatically attaches to any `<form>` tag. It keeps track of the form's value and its validity. On form submission, it aggregates all form control values, accessible via its `value` property.
  - `NgModel`: This directive creates a `FormControl` instance from a domain model and binds it to a form control element. It ensures real-time synchronization between the UI and the component's model.
- `#form="ngForm"`: This creates a local template variable named `form`, which you can use to access the `NgForm` directive instance. This allows you to use its properties and methods elsewhere in the template.
- `onSubmit(form: NgForm)`: This is a method that gets called when the form is submitted. It logs the value of the form to the console.

Reactive forms

In **reactive forms**, you create and manage form control objects within the component class. They are known for their robustness, scalability, reusability, and testability, making them a preferred choice for complex scenarios and large forms. Here is how we would recreate *Figure 5.1* using the reactive forms method:

```
import { CommonModule } from '@angular/common'
import { Component } from '@angular/core'
import { FormControl, FormGroup, ReactiveFormsModule } from '@angular/
forms'

@Component({
  standalone: true,
  imports: [CommonModule, ReactiveFormsModule],
```

```
    template: `
        <h2>Reactive Forms</h2>

        <form [formGroup]="profileForm" (ngSubmit)="onSubmit()">
            <input type="text" formControlName="first" />
            <button type="submit">Submit</button>
        </form>

        <p>Value: {{ profileForm.value | json }}</p>
    `
})
export default class ReactiveFormsComponent {
    profileForm = new FormGroup({
        first: new FormControl(''),
    })

    onSubmit() {
        console.log(this.profileForm.value) // { first: '' }
    }
}
```

Here's a breakdown of important parts of the code:

- `ReactiveFormsModule`, `FormControl`, and `FormGroup`: These are used when you want to use reactive forms in your Angular application.
- `[formGroup]="profileForm"`: This is used to bind the `FormGroup` instance defined in the component class to a form element in the DOM.
- `formControlName`: This directive is used to link the input element to the `first` form control.
- `onSubmit()`: This is a method that gets called when the form is submitted. In the example, the submit function gets the value of `profileForm` and prints the form value to the browser console.

Enhancing Angular forms with PrimeNG input components

Input components and form controls are the backbone of any interactive application. They allow users to interact with the application, input data, and make choices. Without these elements, an application would be a static entity, incapable of interacting with its users in an aesthetic way.

PrimeNG steps in to enhance the standard input elements. It offers a rich set of components with a unified approach to handling user input. PrimeNG's components are designed to be easy to use and integrate into your Angular applications, while also providing a high degree of customization.

For instance, let's take a look at how PrimeNG enhances the previous reactive forms field:

```
import { ButtonModule } from 'primeng/button'
import { InputTextModule } from 'primeng/inputtext'

@Component({
   standalone: true,
   imports: [CommonModule, ReactiveFormsModule, ButtonModule,
InputTextModule],
   template: `
      <h2>PrimeNg Reactive Forms</h2>

      <form [formGroup]="profileForm" (ngSubmit)="onSubmit()">
         <input pInputText formControlName="first" />
         <button pButton type="submit">Submit</button>
      </form>

      <p>Value: {{ profileForm.value | json }}</p>
   `,
})
```

In this code, `pInputText` and `pButton` are PrimeNG directives that improve the standard input field and button and enhance them with additional features such as theming, styles, and more. Compared to *Figure 5.1*, you can see the new form in *Figure 5.2*:

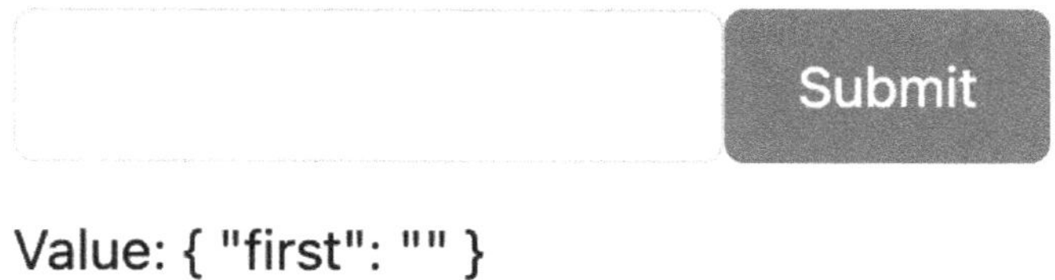

Figure 5.2 – Angular form with PrimeNG styles

In conclusion, understanding input components and form controls is crucial for building interactive and user-friendly applications. Angular provides powerful tools for handling user input, and PrimeNG enhances these tools with a rich set of customizable components. In the following sections, we'll delve deeper into how to use these components and controls in your Angular applications.

Working with text inputs, checkboxes, and radio buttons

As we delve deeper into the practical aspects of Angular and PrimeNG, we'll focus on the implementation of text inputs, checkboxes, and radio buttons. These form controls are fundamental to any application, enabling users to interact with the application and provide necessary data.

Let's take a look at this contact form component:

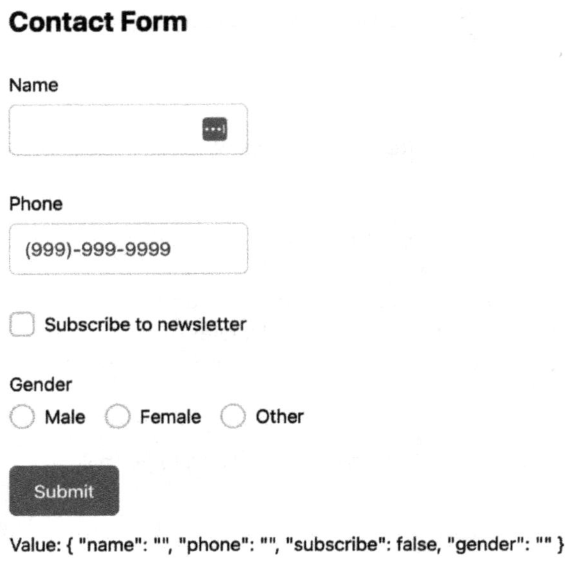

Figure 5.3 – Sample contact form

The contact form utilizes various PrimeNG components such as InputText, InputMask, Checkbox, and RadioButton. Let's take a look at each part.

InputText

The **pInputText** directive is used to enhance existing text input fields. To use the `pInputText` directive in your Angular project, you first need to import the `InputTextModule` module from PrimeNG. You can do this by adding the following `import` statement to your component file:

```
import { InputTextModule } from 'primeng/inputtext'
```

Next, you can use the `pInputText` directive in your template file to create a text input field. Here's the example that we used in the contact form from *Figure 5.3*:

```
<label for="name">Name</label>
<input pInputText id="name" type="text" formControlName="name" />
```

Let's break down the code:

- `<label for="name">Name</label>`: This is a standard HTML label element. The `for` attribute associates the label with the input field that has an ID of `name`.
- `pInputText`: This directive tells Angular to apply the PrimeNG text input functionality and styling to this input field.
- `id="name"`: This attribute sets the ID of the input field, which is used to associate it with the label.
- `type="text"`: This attribute sets the type of the input field. In this case, it's a text field.
- `formControlName="name"`: This attribute is part of Angular's reactive forms module. It binds the input field to a `FormControl` named `name` in the component class.

InputMask

As we delve deeper into PrimeNG's form controls, let's turn our attention to a component that offers a more controlled input experience: **p-inputMask**. This component is designed to handle inputs that follow a specific format, such as phone numbers, dates, or social security numbers.

To use the `p-inputMask` component in your Angular project, you first need to import the `InputMaskModule` module from PrimeNG. You can do this by adding the following `import` statement to your component file:

```
import { InputMaskModule } from 'primeng/inputmask'
```

Here's the example that we used in the contact form in *Figure 5.3*:

```
<label for="phone">Phone</label>
<p-inputMask
   id="phone"
   mask="(999)-999-9999"
   formControlName="phone"
   placeholder="(999)-999-9999"
/>
```

In this example, we're creating an input field for a phone number. The `p-inputMask` component is used as an input field, enforcing the phone number format. Let's break down the code further:

- `<label for="phone">Phone</label>`: This is a standard HTML label element. The `for` attribute associates the label with the input field that has an ID of `phone`.
- `p-inputMask`: This PrimeNG component is used to apply the PrimeNG styles to the input field and to define the input format for the phone number.

- `mask="(999)-999-9999"`: This attribute sets the input pattern for a specific field. In this case, the mask consists of placeholders represented by the character 9, which indicates that only numeric characters can be entered in those positions. By applying this mask, users are restricted to inputting numbers in the designated places, ensuring data consistency and accuracy.
- `formControlName="phone"`: This attribute is part of Angular's reactive forms module. It binds the input field to a `FormControl` named `phone` in the component class.

Checkbox

As we continue to explore PrimeNG's form controls, let's focus on a component that allows users to make binary choices: **p-checkbox**. This component is used to create checkboxes, which let users select one or more options from a set.

To use the `p-checkbox` component in your Angular project, you first need to import the `CheckboxModule` module from PrimeNG. You can do this by adding the following `import` statement to your component file:

```
import { CheckboxModule } from 'primeng/checkbox'
```

Here's the example that we used in the previous contact form in *Figure 5.3*:

```
<p-checkbox
    formControlName="subscribe"
    [binary]="true"
    label="Subscribe to newsletter"
/>
```

Let's break down the code:

- `p-checkbox`: This PrimeNG component is used to apply the PrimeNG styles to the checkbox field.
- `formControlName="subscribe"`: This attribute is part of Angular's reactive forms module. It binds the checkbox to a `FormControl` named `subscribe` in the component class.
- `[binary]="true"`: This attribute sets the checkbox's value to either `true` or `false`. If the checkbox is checked, the value is `true`; otherwise, it's `false`.
- `label="Subscribe to newsletter"`: This attribute sets the label displayed next to the checkbox.

RadioButton

PrimeNG's `p-radioButton` is a UI component that can be used to create radio button inputs in Angular forms. It is a useful component that allows users to make a single choice from a set of mutually exclusive options, such as in surveys, preference selection, form validation, filtering, sorting, and step-by-step processes.

To use the `p-radioButton` component in your Angular project, you first need to import the `RadioButtonModule` module from PrimeNG. You can do this by adding the following `import` statement to your component file:

```
import { RadioButtonModule } from 'primeng/radiobutton'
```

Here's the example that we used in the previous contact form in *Figure 5.3*:

```
<p-radioButton
   ngFor="let gender of genders"
   name="gender"
   value="{{ gender.value }}"
   label="{{ gender.name }}"
   formControlName="gender"
/>
```

Let's break down the code:

- `p-radioButton`: This is where we define our PrimeNG radio buttons.
- `*ngFor="let gender of genders`": This is Angular's built-in directive for rendering a list. It creates a new radio button for each gender in the `genders` array.
- `name="gender"`: This attribute sets the name of the radio button group. All radio buttons with the same name belong to the same group, and only one can be selected at a time.
- `value="{{ gender.value }}"`: This attribute sets the value of the radio button. It's bound to the `value` property of the current `gender` object.
- `label="{{ gender.name }}"`: This attribute sets the label displayed next to the radio button. It's bound to the `name` property of the current `gender` object.
- `formControlName="gender"`: This attribute is part of Angular's reactive forms module. It binds the group of radio buttons to a `FormControl` named `gender` in the component class.

We have learned about some fundamental form controls that allow users to interact with your application. In the next section, we will go over more complex components such as dropdowns, multi-selects, and date pickers.

Using dropdowns, multi-selects, and date pickers

As we continue our exploration of PrimeNG's form controls, we're about to venture into the realm of more complex components: dropdowns, multi-selects, and date pickers. These components offer a higher level of interactivity and are essential for many types of applications.

Dropdown

The **Dropdown** component (also known as Select) is one of the most commonly used form elements in web applications, allowing users to choose a single option from a list of options.

PrimeNG Dropdown offers an intuitive and interactive way to select an option from a dropdown list. It enhances the user experience by providing a wide range of options, including single and multiple selection, filtering, custom templates, lazy loading, and support for keyboard navigation. With its customizable styling and seamless integration with Angular, PrimeNG Dropdown is a powerful tool for creating enhanced dropdown functionality in Angular applications.

To get started with the PrimeNG Dropdown, first import the `DropdownModule` in your component:

```
import { DropdownModule } from 'primeng/dropdown'
```

Here's a basic example of how to use the `Dropdown` component using the template-driven approach:

```
<p-dropdown
        [options]="cities"
        ngModel
        optionLabel="name"
        name="city"
/>
```

In this example, `cities` is an array of options, and `ngModel` is a two-way binding property that holds the selected option. The `optionLabel` attribute configures the Dropdown to display the name property of each option as the label. Here is the result in the browser:

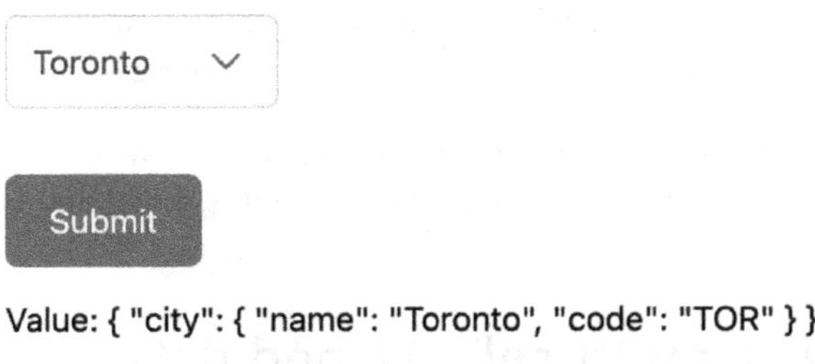

Figure 5.4 – Sample dropdown component

> **Note**
>
> When options are simple primitive values such as a string array, you don't have to specify `optionLabel` and `optionValue` in the component.

The Dropdown component is packed with features that make it flexible and adaptable to various use cases. Let's have a look at those options.

Filtering

One of the standout features of the Dropdown component is its built-in filtering capability. This feature allows users to narrow down the options by typing into the dropdown, making it easier to find the desired option in a long list. To enable filtering, we simply set the `filter` attribute to `true`:

```
<p-dropdown
    [options]="cities"
    ngModel
    optionLabel="name"
    name="cityWithFilter"
    [filter]="true"
/>
```

After adding the `filter` attribute to the Dropdown component, you will have the option to search in the dropdown. Here, I searched for `van`, which shows the result **Vancouver**:

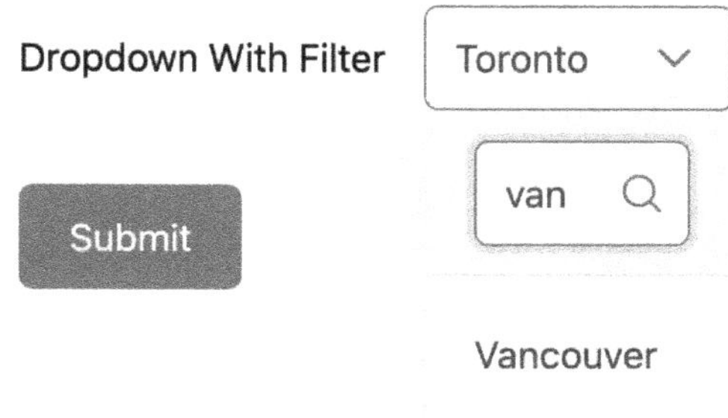

Figure 5.5 – Dropdown With Filter

Grouping

For better organization of options, the Dropdown component supports grouping. We can group options under certain categories by using the [group] attribute:

```
groupedCities = [
    {
        label: 'Canada',
        value: 'ca',
        items: [
            { label: 'Vancouver', value: 'Vancouver' },
            { label: 'Toronto', value: 'Toronto' },
```

```
            { label: 'Montreal', value: 'Montreal' },
            { label: 'Ottawa', value: 'Ottawa' },
        ],
    },
    ...
]

...

<p-dropdown
    [options]="groupedCities"
    ngModel
    name="cityWithGroup"
    placeholder="Select a City"
    [group]="true"
>
    <ng-template let-group pTemplate="group">
        <div>
            <span class="pi pi-map-marker"></span>
            <span>{{ group.label }}</span>
        </div>
    </ng-template>
</p-dropdown>
```

In this example, the cities are grouped according to a property specified in the `groupedCities` array. The `ng-template let-group pTemplate="group"` code defines a template for the group header with a map maker icon on the left. Let's have a look at the result:

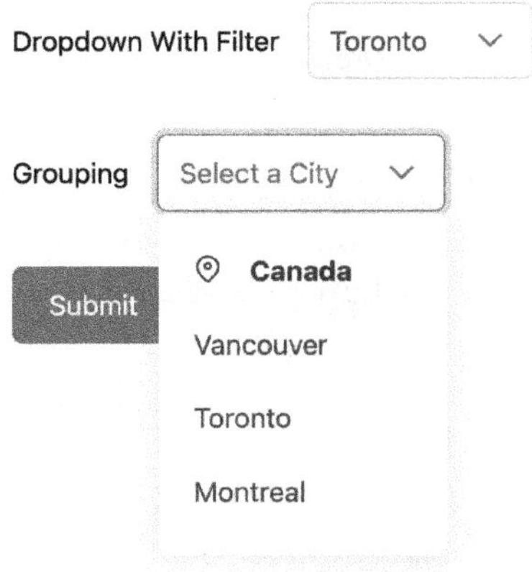

Figure 5.6 – Sample dropdown with Grouping

Templating

The Dropdown component also supports templating, which means we can customize how the options are displayed. Let's take a look at the following example:

```
<p-dropdown
   [options]="cities"
   ngModel
   optionLabel="name"
   name="cityWithTemplate"
   [showClear]="true"
   placeholder="Select a City"
>
   <ng-template pTemplate="selectedItem">
      <div *ngIf="form.value.cityWithTemplate">
         <span class="pi pi-map-marker"></span>
         <span>{{ form.value.cityWithTemplate.name }}</span>
      </div>
   </ng-template>

   <ng-template let-city pTemplate="item">
      <div>
         <span class="pi pi-map-marker"></span>
         <span>{{ city.name }}</span>
      </div>
   </ng-template>
</p-dropdown>
```

In this example, we can see that there are two `ng-template` elements for supporting how we style the option, and the selected value:

- `<ng-template pTemplate="selectedItem">`: This defines a template that is used to customize the rendering of the selected item in a dropdown component. It shows the `name` property of the selected city in the model.
- `<ng-template let-city pTemplate="item">`: This is a template for each item in the option list.

Let's have a look at the result:

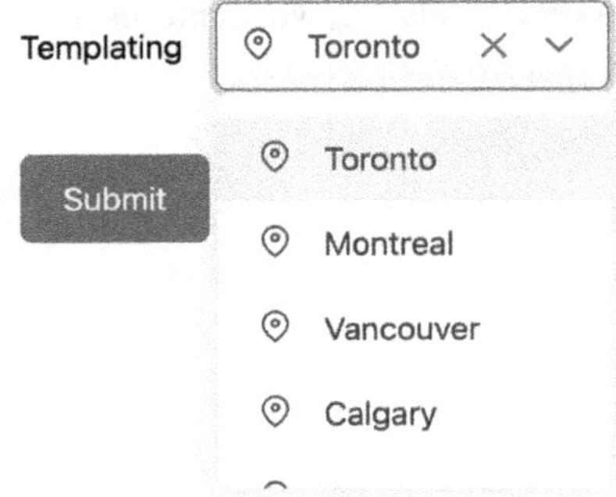

Figure 5.7 – Sample dropdown with Templating

We can see the styling for the selected city with a map maker icon on the left. The dropdown also has the ability to clear the selected item with a `[showClear]="true"` option in the component.

Working with events

PrimeNG's Dropdown component provides several events that developers can utilize to enhance the functionality and interactivity of their applications:

- `onChange`: This is emitted when the value of the dropdown changes
- `onFilter`: This is triggered when data is filtered in the dropdown
- `onFocus`: This is fired when the dropdown gains focus
- `onBlur`: This is invoked when the dropdown loses focus
- `onClick`: This is called when the component is clicked
- `onShow`: This is triggered when the dropdown overlay becomes visible
- `onHide`: This is fired when the dropdown overlay becomes hidden
- `onClear`: This is invoked when the dropdown value is cleared
- `onLazyLoad`: This is called in lazy mode to load new data

Here's an example of how you can use the `onChange` event of the PrimeNG Dropdown when a user selects an option:

```
<p-dropdown
    [options]="cities"
    ngModel
    optionLabel="name"
```

```
    (onChange)="onCityChange($event.value)"
    name="cityWithEvents"
/>

...

onCityChange(value: { name: string; code: string }) {
    alert(`You have selected: ${value.name}`)
}
```

In this example, the `onCityChange` method is called whenever the selected city changes and shows an alert in the browser.

To wrap up, the Dropdown component in PrimeNG is a powerful tool that offers a wide range of features, from basic selection to advanced features such as filtering, grouping, and templating. In the next section, we will delve into the MultiSelect component.

> **Note**
>
> There are more features and configurations that you can do with the Dropdown component. Please visit `https://primeng.org/dropdown` for updated documentation.

MultiSelect

The **MultiSelect** component is a form of select input that allows the user to choose multiple options from a dropdown list, which comes in handy when you need to provide a list of options and allow users to select more than one. For instance, in a survey form, you might ask users to select all the programming languages they are proficient in. The MultiSelect component would be an ideal choice for this scenario.

To get started with the PrimeNG `MultiSelect` component, first import the `MultiSelectModule` in your component:

```
import { MultiSelectModule } from 'primeng/multiselect'
```

Here's a basic example of how to use the `MultiSelect` component using the template-driven approach:

```
cities = [
    { name: 'Toronto', code: 'TOR' },
    { name: 'Montreal', code: 'MTL' },
    { name: 'Vancouver', code: 'VAN' },
    { name: 'Calgary', code: 'CGY' },
    { name: 'Ottawa', code: 'OTT' },
    { name: 'Edmonton', code: 'EDM' },
```

```
    { name: 'Quebec City', code: 'QUE' },
    { name: 'Winnipeg', code: 'WIN' },
    { name: 'Hamilton', code: 'HAM' },
    { name: 'Kitchener', code: 'KIT' },
]

...

<p-multiSelect
    [options]="cities"
    ngModel
    optionLabel="name"
    name="city"
/>
```

In the code snippet, `cities` is an array of options that you want to display in the `MultiSelect` list, while `optionLabel` indicates the dropdown to display the `name` property of each option as the label. Let's look at the result:

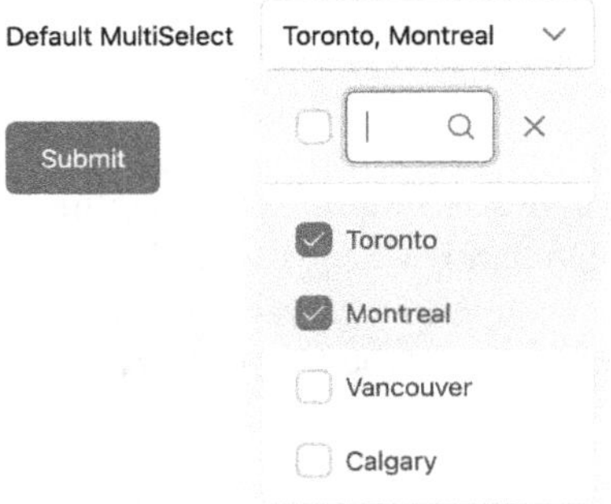

Figure 5.8 – Sample of MultiSelect

In the preceding screenshot, we have a basic `MultiSelect` component, in which **Toronto** and **Montreal** are selected.

The `MultiSelect` component comes with several features that make it a powerful tool for creating interactive forms. Let's take a look at some of them.

Filtering

One of the standout features of the `MultiSelect` component is its built-in filtering capability, which allows users to narrow down the options in the dropdown list by typing into a search box. This is particularly useful when dealing with a large number of options, for example:

```
<p-multiSelect [options]="cities" [filter]="true" />
```

By setting the `filter` attribute to `true`, you enable the filtering feature. Now, when a user clicks on the **MultiSelect** box, they will see a search box at the top of the dropdown list.

Grouping

For better organization of options, the `MultiSelect` component also supports grouping. We can group options under certain categories by using the `[group]` attribute:

```
groupedCities = [
    {
        label: 'Canada',
        value: 'ca',
        items: [
            { label: 'Vancouver', value: 'Vancouver' },
            { label: 'Toronto', value: 'Toronto' },
            { label: 'Montreal', value: 'Montreal' },
            { label: 'Ottawa', value: 'Ottawa' },
        ],
    },
    ...
]

...

<p-multiSelect
    [options]="groupedCities"
    [group]="true"
    ngModel
    name="cityWithGroup"
    defaultLabel="Select a City"
>
    <ng-template let-group pTemplate="group">
        <div>
            <span class="pi pi-map-marker"></span>
            <span>{{ group.label }}</span>
        </div>
    </ng-template>
</p-multiSelect>
```

In this example, the cities are grouped according to a property specified in the `groupedCities` array. The code `ng-template let-group pTemplate="group"` defines a template for the group header with a map maker icon on the left. Let's have a look at the result:

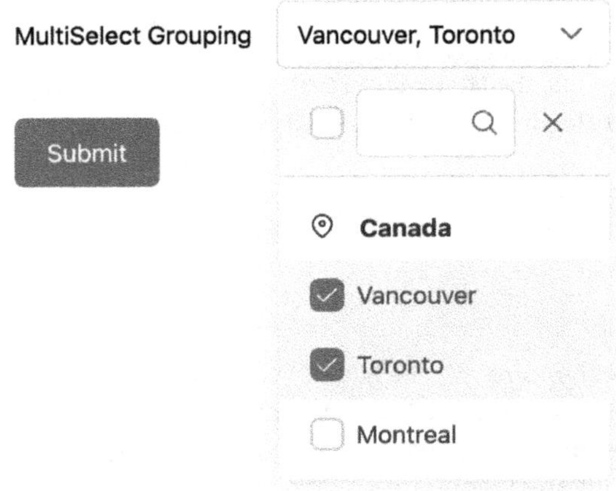

Figure 5.9 – Sample MultiSelect with grouping

Templating

The `MultiSelect` component also supports templating, which allows you to customize the appearance of the options in the dropdown list and the selected items:

```
<p-multiSelect
    [options]="cities"
    ngModel
    defaultLabel="Select a Country"
    name="cityWithTemplating"
    optionLabel="name"
>
    <ng-template let-cities pTemplate="selectedItems">
        <div *ngFor="let city of cities">
            <span class="pi pi-map-marker"></span>
            <span>{{ city.name }}</span>
        </div>

        <div *ngIf="cities?.length === 0">Select Cities</div>
    </ng-template>
```

```
    <ng-template let-city pTemplate="item">
        <div>
            <span class="pi pi-map-marker"></span>
            <span>{{ city.name }}</span>
        </div>
    </ng-template>
</p-multiSelect>
```

In this code snippet, we use the `ng-template` element to define a template for the items in the selection list. The `let-cities` attribute in `<ng-template let-cities pTemplate="selectedItems">` creates a local variable called `cities` that holds the current selected cities. The `<ng-template let-city pTemplate="item">` attribute tells PrimeNG to use this template for the items in the dropdown list.

> **Note**
>
> When checking the PrimeNG documentation, under the **API** tab, you will discover a comprehensive list of templates. For instance, if you are interested in the templates specifically designed for the `MultiSelect` component, you can find the complete list here: `https://primeng.org/multiselect#api.multiselect.templates`.

The PrimeNG `MultiSelect` component is a versatile tool that can enhance the user experience of your forms. In the next section, we will look into the `Calendar` component.

Calendar

PrimeNG **Calendar** is a date picker component that allows users to select dates, times, or both. It's highly customizable, supporting various formats and modes such as inline, button, icon, and input styles.

To get started with the PrimeNG `Calendar` component, first import `CalendarModule` in your component:

```
import { CalendarModule } from 'primeng/calendar'
```

PrimeNG provides the `p-calendar` component for creating date pickers. Here's a basic example:

```
<p-calendar ngModel name="calendar" />
```

In this example, `name="calendar"` is a form control in your form and will hold the selected date:

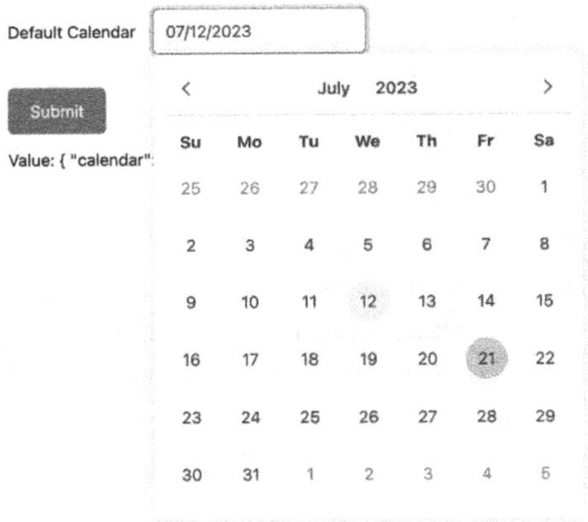

Figure 5.10 – Sample calendar

The PrimeNG Calendar component isn't just a simple date picker. It has a wealth of advanced features that can cater to almost any requirement you might have.

Format

The selected date has the default format of `dd/mm/yyyy`. However, you can update the format of the selected date by using the `dateFormat` property. Here is the list of the available options for `dateFormat`:

- `d`: Day of month (no leading zero).
- `dd`: Day of month (two digits).
- `o`: Day of the year (no leading zeros).
- `oo`: Day of the year (three digits).
- `D`: Day name, short.
- `DD`: Day name, long.
- `m`: Month of year (no leading zero).
- `mm`: Month of year (two digits).
- `M`: Month name, short.
- `MM`: Month name, long.
- `y`: Year (two digits).

- `yy`: Year (four digits).
- `@`: Unix timestamp (milliseconds (ms) – since 01/01/1970).
- `!`: Windows ticks (100ns (nanoseconds) – since 01/01/0001).
- `'...'`: Literal text allows you to insert any text as it is in the date string. For example, `'Day: 'dd 'Month: 'MM 'Year: 'yy` will turn into `Day: 08 Month: August Year: 2023`.
- `''`: If you want to display a single quote within your date string, you must use two single quotes together, for example, `'Today''s date is 'dd/MM/yy` will turn into `Today's date is 08/August/2023`.
- Any other characters that are not recognized as part of the date format pattern will be treated as literal text and will be displayed as is. For example, if you use the format string `dd+MM+yy`, a date might display as `16+August+2023`.

Here is an example of changing the format of the selected date:

```
<p-calendar
    ngModel
    name="calendarWithFormat"
    dateFormat="dd-mm-yy"
/>
```

In this example, `dateFormat="dd-mm-yy"` indicates the new format of the selected date. Here is the result:

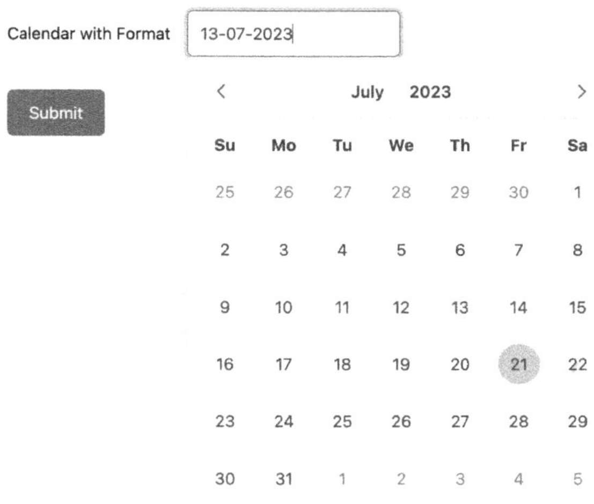

Figure 5.11 – Sample calendar

After selecting the date, you can see the format of the selected date changes to `13-07-2023`.

Range selection

To enable range selection in the PrimeNG Calendar component, you can use the `selectionMode` attribute along with the `ngModel` directive. The `selectionMode` attribute allows you to specify the mode of selection, including `"single"`, `"multiple"`, and `"range"`. Here's how you can enable range selection:

```
<p-calendar
   ngModel
   name="calendarDateRange"
   selectionMode="range"
/>
```

Here, `calendarDateRange` is an array that will hold the start and end dates of the selected range.

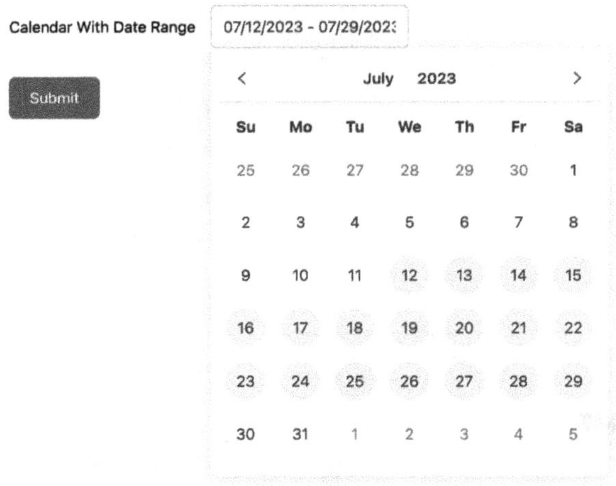

Figure 5.12 – Sample calendar with date range

PrimeNG Calendar is a powerful, flexible, and customizable date picker component for Angular applications. It also offers extensive customization options, allowing you to tailor its appearance and behavior to your needs.

> **Note**
>
> In this section, we've only scratched the surface of what the PrimeNG `Calendar` can do. We encourage you to explore its documentation to learn more about its features and capabilities: `https://primeng.org/calendar`.

In this section, we went through some important input components. In the next section, let's go through validation and input handling techniques.

Implementing form validation

As we venture deeper into the realm of form controls, we come across a critical aspect of any form—**validation**. Ensuring the data entered by users is valid and as expected is paramount to maintaining data integrity and providing a seamless user experience. It's not just about checking whether a field is empty or not; it's about ensuring the data is in the right format, within certain limits, or meets any other criteria you set. In this section, we'll explore how to implement form validation using Angular and PrimeNG.

Understanding Angular form states

Before diving into the details of Angular form validation, let's first discuss some important concepts related to form validation. In Angular, a form is represented by the `FormGroup` class, which contains an organized collection of form controls such as input fields, checkboxes, and dropdowns.

Angular form validation revolves around the state of the form controls. There are several states that a form control can have:

- `pristine`: A form control is considered pristine if the user hasn't interacted with it yet. It means that the control's value has not been changed since it was initialized.
- `dirty`: A form control becomes dirty once the user interacts with it and modifies its value.
- `touched`: A form control is marked as touched when the user focuses on it and then moves away, indicating that they have interacted with the control.
- `untouched`: The opposite of touched, an untouched form control means that the user hasn't interacted with it yet. The difference between `pristine` and `untouched` is that `untouched` refers to a form control that has not been interacted with by the user, while `pristine` indicates a form control that has not been modified.
- `valid`: A form control is considered valid if it satisfies all the validation rules defined for it.
- `invalid`: If a form control fails to satisfy any of the validation rules, it is marked as invalid.

These states play a crucial role in form validation, as they help determine the visual feedback to provide to the user and enable or disable form submission based on the validity of the form controls.

Built-in Angular form validation

Angular provides a set of built-in validators that cover common form validation scenarios. These validators can be used on form controls to ensure that the user input meets specific requirements. Here are a few examples:

- `required`: The `required` validator ensures that a form control has a non-empty value
- `minLength` and `maxLength`: These validators validate the minimum and maximum length of a form control's value, respectively
- `pattern`: The `pattern` validator allows you to specify a regular expression pattern that the form control's value must match

To apply these validators to a form control, you need to associate them with the control in the component code. For example, if you have an input field for the user's name that must be filled in and has a minimum length of four characters, you can define the form with validators like this (in a template-driven form):

```
<input
    name="first"
    ngModel
    required
    minlength="4"
    #name="ngModel"
/>

<ng-container *ngIf="name.invalid && (name.dirty || name.touched)">
    <div *ngIf="name.errors?.['required']">
        This field is required
    </div>
    <div *ngIf="name.errors?.['minlength']">
        Name must be at least 4 characters long.
    </div>
</ng-container>
```

Let's break down the example:

- `required`: This indicates that the input must have a value before the form can be submitted.
- `minlength="4"`: This specifies that the input value must be at least four characters long.
- `#name="ngModel"`: This creates a local template variable called `name`. This allows us to access the properties and methods of the `NgModel` directive within the template.

- `*ngIf="name.invalid && (name.dirty || name.touched)"`: The `*ngIf` directive checks if the input is invalid (`name.invalid`) and if the input has been interacted with (`name.dirty` or `name.touched`). If both conditions are true, the content inside `<ng-container>` will be displayed.
- `*ngIf="name.errors?.['required']"`: This displays an error message if the input is missing a value (because of the `required` attribute).
- `*ngIf="name.errors?.['minlength']"`: This displays an error message if the input value is less than four characters long (because of the `minlength="4"` attribute).

Now let's look at validators in reactive forms:

```
contactForm = this.formBuilder.group({
  name: ['', [Validators.required, Validators.minLength(4)]],
})

...
<input pInputText id="name" type="text" formControlName="name" />

<ng-container *ngIf="contactForm.controls.name as name">
  <div *ngIf="name.dirty && name.hasError('required')">
    This field is required
  </div>
  <div *ngIf="name.dirty && name.hasError('minlength')">
    Name must be at least 4 characters long.
  </div>
</ng-container>
```

Let's break down the code:

- `name: ['', [Validators.required, Validators.minLength(4)]]`: This initializes a form control called name with an empty string as its default value. The array that follows specifies the validation rules for this control:
  - `Validators.required`: This validator ensures that the control has a non-empty value
  - `Validators.minLength(4)`: This validator ensures that the control's value is at least four characters long
- `*ngIf="contactForm.controls.name as name"`: This checks if the name control exists in `contactForm` and assigns it to a local template variable, name
- `*ngIf="name.dirty && name.hasError('required')`: This displays an error message if the name control has been interacted with (`name.dirty`) and if it's missing a value (`name.hasError('required')`)

- `*ngIf="name.dirty && name.hasError('minlength')"`: This displays an error message if the `name` control has been interacted with and if its value is less than four characters long

Based on the logic just explained, we can create the entire form with proper validation. *Figure 5.13* shows a contact form with an invalid state:

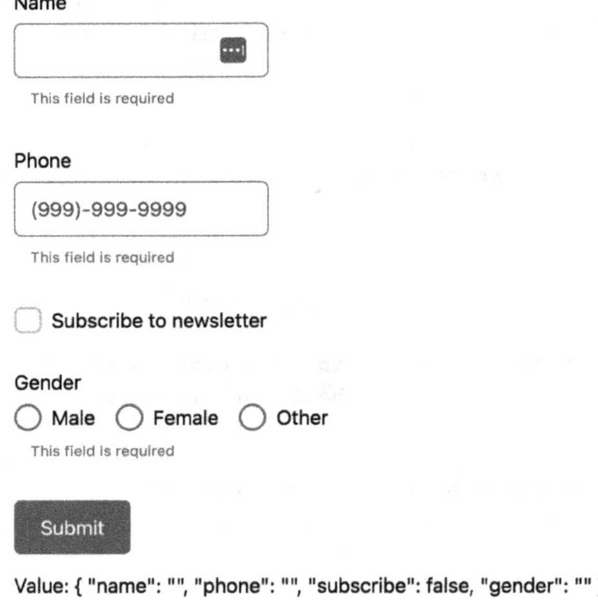

Figure 5.13 – Sample form validation

As we can see, the error state indicates that the form requires all of the fields to be filled out before the information can be submitted.

Crafting custom form validation

While Angular provides a range of built-in validators, there are times when we need to add something specific to our application. Thankfully, Angular allows us to create custom form validators.

Here's a simple example of creating a custom validator that checks whether the name is invalid or not:

```
import { AbstractControl, ValidationErrors, ValidatorFn } from '@
angular/forms'

export const invalidNameValidator = (nameRe: RegExp): ValidatorFn => {
```

```
    return (control: AbstractControl): ValidationErrors | null => {
      const invalid = nameRe.test(control.value)
      return invalid ? { invalidName: { value: control.value } } :
null
    }
  }
```

In the code example, the main function, `invalidNameValidator`, takes a regular expression (`nameRe`) as its argument and returns a custom validator function. This custom validator function checks whether a form control's value matches the provided regular expression. If the value doesn't match the pattern, meaning that it's valid, the function returns `null`, indicating that there are no validation errors.

> **Note**
>
> Custom validators return a key-value pair if the validation fails. The key is a string of your choosing that describes the error, and the value is the value of the control.

To use this validator, we just need to add it to the existing validators:

```
contactForm = this.formBuilder.group({
    name: [
      '',
      [
        Validators.required,
        Validators.minLength(4),
        // custom validator
        invalidNameValidator(/test/i),
      ],
    ],
    ...
})

----

<div *ngIf="name.dirty && name.hasError('invalidName')">
    Name cannot be "{{ name.errors?.['invalidName'].value }}".
</div>
```

Here is the code breakdown:

- `invalidNameValidator(/test/i)`: This is a custom validator (as we discussed in the previous section). It checks whether the form control's value matches the regular expression `/test/i`. If the value matches, the validator will flag it as invalid.

- `*ngIf="name.dirty && name.hasError('invalidName')"`: This displays an error message if the `name` control has been interacted with and if the name is invalid.

Let's look at the result:

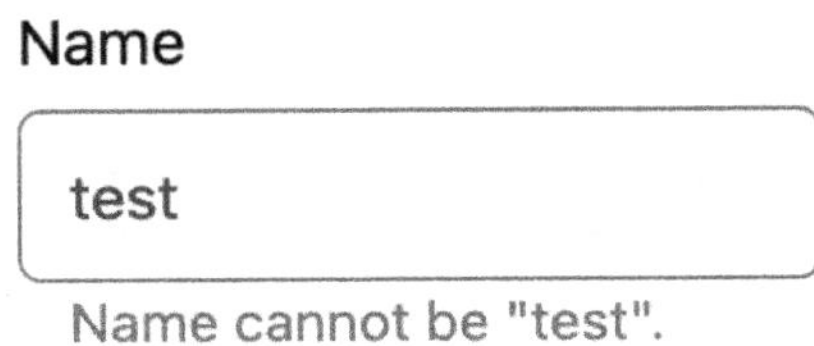

Figure 5.14 – Custom form validation

PrimeNG and form validation

PrimeNG complements Angular's form validation by providing visual feedback for validation errors. It includes a variety of CSS classes that can be used to highlight invalid fields and display error messages.

Here's an example of how to use PrimeNG to display validation errors:

```
<label for="name">Name</label>
<input pInputText id="name" type="text" formControlName="name" />
<ng-container *ngIf="contactForm.controls.name as name">
   <small
      class="p-error"
        ngIf="name.dirty && name.hasError('required')"
   >
      This field is required
   </small>
   <small
      class="p-error"
        ngIf="name.dirty && name.hasError('minlength')"
   >
      The name is too short
   </small>
</ng-container>
```

In this example, if the `name` field is invalid and `dirty`, the `p-error` class is added to the field, and the error message is displayed.

Form validation is a critical aspect of any application that involves user input. It helps ensure data integrity and enhances the user experience. Angular provides robust support for form validation, and PrimeNG complements this by providing visual feedback for validation errors.

In this section, we've explored how to implement form validation using Angular and PrimeNG. However, this is just the tip of the iceberg. Both Angular and PrimeNG offer a wealth of features for form validation that can cater to almost any requirement We encourage you to explore Angular documentation to learn more: `https://angular.io/guide/form-validation`

Summary

In this chapter, we delved into the world of input components and form controls, exploring their usage in Angular applications with PrimeNG. We started by understanding their importance in building interactive applications, setting the foundation for the practical aspects that followed.

We then navigated through the implementation of basic form controls such as text inputs, checkboxes, and radio buttons, and moved on to more complex ones such as dropdowns, multi-selects, and date pickers. Each section was enriched with code examples, demonstrating the integration of these components into our applications.

Toward the end, we tackled the crucial topic of form validation. We emphasized the importance of validating user input for enhancing user experience and ensuring data integrity, guiding you through the process with Angular and PrimeNG.

As we look forward to the next chapter, on data tables and other displaying components, remember that the knowledge gained here is vital for any web developer. Form controls are fundamental to web applications, and understanding their effective use is key to creating user-friendly applications.

6
Working with Table, List, and Card Components

As we continue our journey with PrimeNG and Angular, we find ourselves in the realm of data display components. In this chapter, we will focus on three key components: the data table, list, and card components. These components are the workhorses of any application, responsible for presenting data to users in a clear, concise, and user-friendly manner. They are the bridge between the raw data in our applications and the polished, interactive interfaces that our users interact with.

The objective of this chapter is to provide you with the necessary knowledge and skills to effectively utilize data tables, lists, and card components, enabling you to present data in the most user-friendly and efficient manner possible. By gaining mastery over these components, you can ensure that users can effortlessly comprehend and interact with the data, ultimately leading to enhanced user engagement and satisfaction.

In this chapter, we will cover the following topics:

- Creating responsive layouts with PrimeFlex
- Introducing data display components
- Working with data table components
- Working with list components
- Working with card components

Technical requirements

This chapter contains various code samples of PrimeNG displaying components. You can find the related source code in the `chapter-06` folder of the following GitHub repository: `https://github.com/PacktPublishing/Next-Level-UI-Development-with-PrimeNG/tree/main/apps/chapter-06`.

Creating responsive layouts with PrimeFlex

PrimeFlex is a lightweight, responsive CSS utility library designed to accompany Prime UI libraries and static web pages. It's a perfect CSS utility companion that empowers web design by providing a collection of pre-built components and utility classes. In this section, we'll explore PrimeFlex and how it's used with PrimeNG, including building layouts, using `Flexbox` and `Grid`, and more.

Integrating PrimeFlex with PrimeNG

PrimeFlex can be easily integrated with PrimeNG by installing it via `npm`:

```
npm install primeflex
```

After the installation process, we'll include the PrimeFlex library in our `styles.scss` file:

```
// styles.scss

@import 'primeflex/primeflex.scss';
```

Once we import `primeflex.scss`, we will be able to create an Angular application consistently, ensuring proper spacing, typography, layout, and all other essential elements.

Let's compare how we build layouts with and without PrimeFlex.

Building layouts without a utility library such as PrimeFlex can be cumbersome. You might find yourself writing repetitive and verbose CSS code, as seen in the following code example:

```
<h1 [routerLink]="['/']">Welcome to chapter-06</h1>

<div class="layout-wrapper">
    <aside>
        <nav>
            <p-menu [model]="items" />
        </nav>
    </aside>
    <main>
        <router-outlet />
    </main>
</div>

....
styles: [
    `
        .layout-wrapper {
            display: flex;
```

```
            gap: 4rem;
            flex-wrap: wrap;
        }
      `,
  ],
```

We apply styles by creating the `.layout-wrapper` class and adding CSS syntax to it. This is a standard CSS implementation that is supported by web browsers. With normal CSS, you write your own styles from scratch.

On the other hand, PrimeFlex simplifies this process by providing utility classes that encapsulate common CSS properties. Let's look at the template with PrimeFlex utility classes:

```
<h1 [routerLink]="['/']">Welcome to chapter-06</h1>

<div class="flex flex-wrap gap-7">
    ...
</div>
```

By utilizing common CSS utility classes, we can streamline our template, such as transforming the `.layout-wrapper` class to `flex flex-wrap gap-7`, eliminating the need for writing custom CSS code.

The usage of PrimeFlex utility classes offers ease of use and consistency within your entire Angular application, while normal CSS provides greater flexibility and customization options, although it requires more effort and expertise to achieve desired styles. Let's examine this in the browser:

```
▼<primengbook-root ng-version="16.0.6">
   <h1 tabindex="0" ng-reflect-router-link="/">Welcome to
   chapter-06</h1>
 ▶<div class="flex flex-wrap gap-7">…</div> flex == $0
 </primengbook-root>
html  body  primengbook-root  div.flex.flex-wrap.gap-7
Styles  Computed  Layout  Event Listeners  DOM Breakpoints  >>
Filter                                   :hov .cls +
.gap-7 {                                 _mixins.scss:274
  gap: ▶ 4rem !important;
}
.flex-wrap {                             _mixins.scss:5
  flex-wrap: wrap !important;
}
.flex {                                  _mixins.scss:5
  display: flex !important;
}
```

Figure 6.1 – PrimeFlex classes example

You can observe that the utility classes are mapped to their respective CSS styles, for example, from `flex` to `display: flex !important;`.

Here are some examples of utility classes related to layout (`Flexbox` and `Grid`) and text in PrimeFlex:

Class	Explanation
`flex`	Enables flexbox layout.
`flex-column`	Arranges flex items in a column.
`grid`	Enables grid layout.
`w-full`	Sets the width to 100%.
`gap-2`	Defines the gap between grid or flex items.
`p-4`	Applies padding of 4 units to all sides.
`justify-content-between`	Justifies content with space between items.
`align-items-center`	Aligns items to the center vertically in a flex container.
`text-center`	Centers text within an element.
`text-uppercase`	Transforms text to uppercase.
`text-bold`	Applies bold styling to text.
`text-italic`	Applies italic styling to text.

Figure 6.2 – Common PrimeFlex utility classes

These classes are part of PrimeFlex's utility-first approach to CSS, providing you with a set of reusable classes that encapsulate common layout and text styling patterns. By using these classes, you can quickly build complex layouts and apply text styling without having to write custom CSS code, leading to a more efficient and maintainable development process. For a complete set of utility classes, please check out the documentation at `https://primeflex.org`.

Using Flexbox in PrimeFlex

PrimeFlex provides a robust and versatile **Flexbox** utility system that allows you to create flexible and responsive layouts. With Flexbox, you can easily distribute and align elements within a container, making it an excellent tool for building modern and dynamic user interfaces.

Creating a flex container

To create a Flexbox layout, you need to designate a container element as a flex container. By applying the `flex` class to the container, you enable the Flexbox behavior, allowing child elements to be flex items:

```
<div class="flex">
   <div class="text-center p-3 bg-primary">Flex Item 1</div>
   <div class="text-center p-3 bg-primary">Flex Item 2</div>
   <div class="text-center p-3 bg-primary">Flex Item 3</div>
</div>
```

In the preceding code example, the `flex` class is applied to the container element, which makes it a flex container. The child elements inside the container automatically become flex items.

This is the base code for *Figure 6.3*, which you will see later.

Applying flex direction

Flexbox also provides four main directions to arrange flex items within a flex container: `row`, `row-reverse`, `column`, and `column-reverse`. The direction is determined by applying one of the following classes to the flex container:

- `flex-row`: Items are laid out in a row, from left to right
- `flex-row-reverse`: Items are laid out in a row, from right to left
- `flex-column`: Items are laid out in a column, from top to bottom
- `flex-column-reverse`: Items are laid out in a column, from bottom to top

This is an example of how we create a flex container with different layouts:

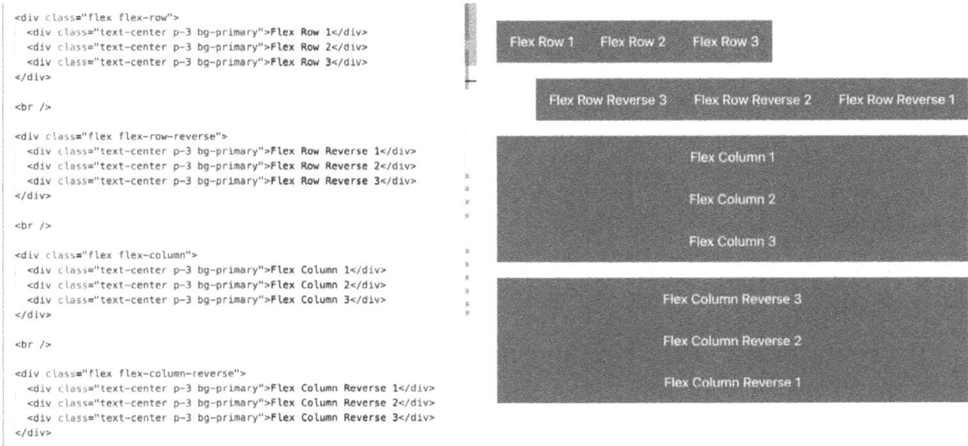

Figure 6.3 – PrimeFlex Flexbox example

In this example, we harness the power of Flexbox by effortlessly creating diverse layouts in either rows or columns simply by applying different classes.

> **Note**
>
> To learn more about Flexbox and how to debug it, you can visit `https://developer.chrome.com/docs/devtools/css/flexbox/`. It's a helpful resource that provides insights and guidance on understanding Flexbox and troubleshooting any issues you may encounter.

Using flex wrap

By default, flex items will try to fit within a single line. However, if there is insufficient space, the flex items will shrink to accommodate. To control the wrapping behavior of flex items, you can use the following classes:

- `flex-wrap`: Items wrap onto multiple lines if necessary
- `flex-wrap-reverse`: Items wrap onto multiple lines if necessary, in reverse order
- `flex-nowrap`: Items remain on a single line

Let's apply wrap functionality to our flex container:

```
<div class="flex flex-wrap">
   ...
</div>
```

In this example, the flex container has the `flex-wrap` class applied, allowing the flex items to wrap onto multiple lines if needed.

Using justify content

Flexbox provides powerful alignment options to position flex items along the main and cross axes. PrimeFlex offers a variety of classes to control alignment:

- `justify-content-end`: This aligns the flex items to the end of the flex container along the main axis
- `justify-content-center`: This centers the flex items along the main axis of the flex container
- `justify-content-between`: This distributes the flex items along the main axis with equal space between them
- `justify-content-around`: This distributes the flex items along the main axis with equal space around them
- `justify-content-evenly`: This distributes the flex items along the main axis with equal space between them, including before the first item and after the last item

These alignment classes can be applied to the flex container or individual flex items:

```
<div class="flex flex-wrap justify-content-evenly">
    ...
</div>
```

In this preceding example, the flex container has the `justify-content-evenly` class applied, which distributes the flex items evenly with equal space around them. The `flex-wrap` class allows items to wrap onto multiple lines if needed.

Taking all of the Flexbox code into consideration, let's have a look at the result:

Figure 6.4 – Flexbox example

You see the element is a flex container with flex items that wrap onto multiple lines if needed. The flex items are evenly distributed along the main axis with equal spacing between them and at the beginning and end of the container.

Using the grid system in PrimeFlex

The grid system in PrimeFlex is another powerful feature for building complex layouts, providing a simple, intuitive syntax for building grids and arranging elements on the page.

The **PrimeFlex grid** follows a 12-column structure. Each row consists of a container, which contains one or more columns. The columns within a row automatically adjust their width based on the available space.

To create a grid layout, you need to wrap your content in a container element with the `grid` class. Inside the container, you can define columns using the `col` class. The columns are specified by adding the `col-{size}` class, where `{size}` represents the number of columns the element should span.

Here is an example of a basic grid structure with two columns:

```
<div class="grid">
   <div class="col-6">
         <div class="text-center p-3 bg-primary">6</div>
   </div>
   <div class="col-6">
         <div class="text-center p-3 bg-primary">6</div>
```

```
    </div>
    <div class="col-6">
            <div class="text-center p-3 bg-primary">6</div>
    </div>
    <div class="col-6">
            <div class="text-center p-3 bg-primary">6</div>
    </div>
</div>
```

In this example, we have a grid container with two columns. Each column spans six columns, resulting in two equal-width columns. Here is the result:

Figure 6.5 – PrimeFlex grid example

The PrimeFlex grid also provides responsive classes that allow you to create different layouts for different screen sizes. You can specify different column sizes based on screen breakpoints using the following syntaxes:

- `sm`: Small screens (576px and above)
- `md`: Medium screens (768px and above)
- `lg`: Large screens (992px and above)
- `xl`: Extra-large screens (1200px and above)

Let's use breakpoints to create a responsive layout:

```
<div class="grid">
    <div class="col-12 md:col-6 lg:col-3">
        <div class="text-center p-3 bg-primary">6</div>
    </div>

    ...
</div>
```

In this example, we have a grid container with two columns. Each column spans the full width by default on small screens (12 columns), half the width on medium screens (six columns), and one-third of the width on large screens (four columns).

PrimeFlex is a valuable tool for web developers, offering a streamlined approach to CSS styling. Its integration with PrimeNG enhances the development experience, providing a consistent and flexible design system. In the following section, we will be introduced to PrimeNG data display components.

Introducing data display components

As we all know, data is the lifeblood of any application, but raw data in itself is not very useful. It's how we present this data to users that truly matters. That's where data display components come into play. They are the tools that transform raw data into meaningful information, providing users with insights and enabling them to interact with the data.

PrimeNG offers a variety of data display components. These components are designed to present data in a clear, concise, and user-friendly manner. They include data tables, lists, cards, and much more. Each of these components has its strengths and use cases, and together, they provide a comprehensive toolkit for data display. Without these components, users would be left with raw data that is difficult to interpret and analyze. This could lead to errors, misunderstandings, and poor decision-making.

Let's take a look at the following PrimeNG data display components and when to use them:

- **Data tables**, for instance, are perfect for displaying large amounts of data in a structured format. They support features such as sorting, filtering, and pagination, making it easy for users to navigate and interact with the data.
- **Lists**, on the other hand, are ideal for displaying a collection of items in a simple and straightforward manner. They are versatile and can be used for a wide range of use cases, from simple lists of text items to complex lists with custom layouts.
- **Cards** are another powerful tool for data display. They are a great way to present a collection of related information in a flexible and extensible format. Cards can contain any type of content, from text and images to buttons and links, and they can be arranged in various ways to create visually appealing layouts.

In the following sections, we'll dive deeper into these components, exploring how to use them in your Angular applications. We'll provide code examples to illustrate their usage and discuss the various options and configurations available for each component.

> **Note**
>
> Remember, the key to effective data display is not just about choosing the right components, but also about using them in the right way. It's about understanding the data, knowing what information is important to the users, and presenting it in a way that is easy to understand and interact with.

Working with data table components

Diving into the world of data display, we find ourselves surrounded by a myriad of components, each with its unique features and capabilities. Among these, PrimeNG tables stand out as a versatile and powerful tool that can transform raw data into meaningful, interactive, and visually appealing information.

To use PrimeNG tables in your Angular project, you first need to import `TableModule` from PrimeNG. You can do this by adding the following `import` statement to your module file:

```
import { TableModule } from 'primeng/table'
```

This section will explore the various features and functionalities of PrimeNG tables, providing you with practical examples and insights to help you leverage their full potential in your applications. In the following examples, we will work with sample product data. Here is the interface of `Product`:

```
interface Product {
    id: number
    name: string
    price: number
    description: string
    quantity: number
    rating: number
    category: string
}
```

So, let's jump into tables.

Creating a basic table

PrimeNG tables require a collection of data to display, along with column components that define how this data should be represented. Here's a simple example:

```
<p-table
    [value]="products"
    [tableStyle]="{ 'min-width': '50rem' }"
>
      <ng-template pTemplate="header">
            <tr>
                  <th>ID</th>
                  <th>Name</th>
                  <th>Category</th>
                  <th>Quantity</th>
```

```
        </tr>
    </ng-template>
    <ng-template pTemplate="body" let-product>
        <tr>
            <td>{{ product.id }}</td>
            <td>{{ product.name }}</td>
            <td>{{ product.category }}</td>
            <td>{{ product.quantity }}</td>
        </tr>
    </ng-template>
</p-table>
```

Let's break down the code snippet:

- `<p-table>`: This is the PrimeNG table component, used to display tabular data.
- `[value]="products"`: This attribute binding is binding the `products` property to the `value` attribute of the `p-table` component. It means that the `products` variable in the component's code is the data source for the table.
- `[tableStyle]="{ 'min-width': '50rem' }"`: This attribute binding is binding an inline CSS style object to the `tableStyle` attribute of the `p-table` component. The provided style object sets the minimum width of the table to `50rem`.
- `<ng-template pTemplate="header">`: This is a template for the header row of the table. It defines the column titles. In this case, the table has four columns: `ID`, `Name`, `Category`, and `Quantity`.
- `<ng-template pTemplate="body" let-product>`: This is a template for the body of the table. It defines how each row of data should be displayed. The `let-product` syntax is used to create a local template variable product that holds the current product object for each row.
- Inside the body template, we have a `<tr>` element for each row, and `<td>` elements for each cell in the row. The `{{ product.id }}`, `{{ product.name }}`, `{{ product.category }}`, and `{{ product.quantity }}` expressions are used to bind the properties of the current product object to the cells.

After that, we will have a basic table, like so:

ID	Name	Category	Quantity
1	Product 1	Category 1	100
2	Product 2	Category 2	50
3	Product 3	Category 1	75
4	Product 4	Category 3	200
5	Product 5	Category 2	150

Figure 6.6 – Basic table

Table with dynamic columns

Columns can be defined dynamically using the `*ngFor` directive. This is particularly useful when the structure of your data is not known in advance or can change dynamically. Here's how you can do it:

```
cols = [
    { field: 'id', header: 'ID' },
    { field: 'name', header: 'Name' },
    { field: 'category', header: 'Category' },
    { field: 'quantity', header: 'Quantity' },
]

...

<p-table
      [columns]="cols"
      [value]="products"
      [tableStyle]="{ 'min-width': '50rem' }"
>
       <ng-template pTemplate="header" let-columns>
              <tr>
                     <th *ngFor="let col of columns">
                            {{ col.header }}
                     </th>
              </tr>
```

```
    </ng-template>
    <ng-template pTemplate="body" let-rowData let-columns="columns">
        <tr>
            <td *ngFor="let col of columns">
                {{ rowData[col.field] }}
            </td>
        </tr>
    </ng-template>
</p-table>
```

In this example, the `columns` attribute is bound to the `cols` array, which contains the definitions of the columns. Each column is an object with a `header` property (the column title) and a `field` property (the property of the data object to bind to). Since the table data is defined in the component, you can provide options for users to select which columns to display, reorder columns, or even dynamically add or remove columns from the table.

Table with sorting

Sorting is a fundamental aspect of data presentation, allowing users to order data in a way that makes sense for their specific tasks. PrimeNG provides built-in functionality for sorting data in tables.

Here is an example of how to enable sorting in a PrimeNG table component:

```
<p-table [value]="products">
    <ng-template pTemplate="header">
        <tr>
            <th pSortableColumn="name">Name
                <p-sortIcon field="name"></p-sortIcon>
            </th>
            <th pSortableColumn="price">Price
                <p-sortIcon field="price"></p-sortIcon>
            </th>
        </tr>
    </ng-template>
    <ng-template pTemplate="body" let-product>
        <tr>
            <td>{{ product.name }}</td>
            <td>{{ product.price }}</td>
        </tr>
    </ng-template>
</p-table>
```

In this example, the `pSortableColumn` directive is used to specify the field by which the data should be sorted when the column header is clicked. The `p-sortIcon` component is used to display an icon indicating the sort order.

By default, clicking the column header once will sort the data in ascending order. Clicking it again will sort the data in descending order. Let's take a look at the result:

Name ↑↓	Price ↑
Product 5	8.99
Product 1	10.99
Product 4	12.99
Product 2	15.99
Product 3	20.99

Figure 6.7 – Table with sorting

You can see that, in the screenshot, the table is sorted by price in ascending order after clicking on the **Price** header.

Table with filtering

Filtering is another powerful feature, allowing users to narrow down the data displayed in the table based on specific criteria. Again, PrimeNG provides built-in functionality for this.

To enable filtering for a column, you need to add some custom templates to the column definition in your table. Here's an example of how to do this:

```
<p-table
    [value]="products"
    [globalFilterFields]="['name', 'price']"
    #dt
>
  <ng-template pTemplate="caption">
    <div class="flex">
      <button pButton label="Clear" class="p-button-outlined"
        icon="pi pi-filter-slash" (click)="dt.clear()"
```

```
            ></button>
            <span class="p-input-icon-left ml-auto">
                <i class="pi pi-search"></i>
                <input pInputText type="text" placeholder="Search keyword"
                    (input)="dt.filterGlobal($event.target.value,
'contains')"            />
            </span>
        </div>
    </ng-template>

    <ng-template pTemplate="header">
        <tr>
            <th>
                <input pInputText type="text" placeholder="Search by name"
                    (input)="dt.filter($event.target.value, 'name',
'contains')"
                />
            </th>
            <th>
                <input pInputText type="text" placeholder="Search by
price"
                    (input)="dt.filter($event.target.value, 'price',
'equals')"
                />
            </th>
        </tr>
    </ng-template>

    <ng-template pTemplate="body" let-product>
        <tr>
            <td>{{ product.name }}</td>
            <td>{{ product.price }}</td>
        </tr>
    </ng-template>

    <ng-template pTemplate="emptymessage">
        <tr>
            <td colspan="7">No products found.</td>
        </tr>
    </ng-template>
</p-table>
```

The previous code is an example of a PrimeNG table with both column and global filtering capabilities. Let's break it down:

- `<p-table>` is a PrimeNG table component, used to display tabular data.
- `[value]="products"` is an attribute binding, binding the `products` property to the `value` attribute of the `<p-table>` component. It means that the `products` variable in the component's code is the data source for the table.
- The `[globalFilterFields]="['name', 'price']"` attribute binding is binding an array of field names (`name` and `price`) to the `globalFilterFields` attribute of the `<p-table>` component. The `globalFilterFields` attribute allows you to specify the fields/columns on which you want to apply the global filter. In this case, the global filter will be applied to the `name` and `price` fields of the table.
- `#dt` is a template reference variable named `dt` that is assigned to the `<p-table>` component. Template reference variables allow you to reference the component in the template code and access its properties and methods if needed.
- The `<ng-template pTemplate="caption">` template contains a button for clearing all filters and an input field for global search. The `dt.clear()` method is called when the button is clicked to clear all filters. The `dt.filterGlobal()` method is called when the user types in the global search input field to filter all rows based on the input value.
- The `<ng-template pTemplate="header">` template contains input fields for each column. The `dt.filter()` method is called when the user types in these input fields to filter the rows based on the input value for the corresponding column.
- The `<ng-template pTemplate="body" let-product>` template defines how each row of data should be displayed. The `let-product` syntax is used to create a local template `product` variable that holds the current product object for each row.
- The `<ng-template pTemplate="emptymessage">` template is displayed when there are no rows to display, either because the `products` array is empty or because no rows match the current filters.

Let's take a look at the final result:

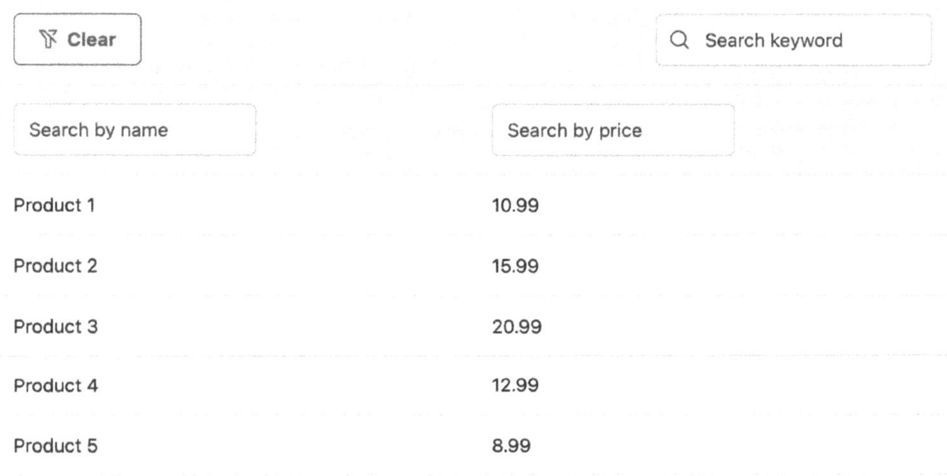

Figure 6.8 – Table with filtering

For example, the user can filter the rows by typing in the input fields in the column headers or the global search input field. The user can also clear all filters by clicking the **Clear** button.

Table with a paginator

When dealing with large datasets, displaying all the data at once can be overwhelming and impractical. **Pagination** is a common solution to this problem, allowing users to view a subset of the data at a time. PrimeNG's table component comes with a built-in paginator that makes implementing this functionality a breeze.

To enable pagination in a PrimeNG table, you simply need to set the `paginator` attribute to `true` and define the `rows` attribute to specify the number of rows per page:

```
<p-table
    [value]="products"
    [paginator]="true"
    [rowsPerPageOptions]="[5,10,20]"
    [rows]="10"
    >
    <!-- Table content -->
</p-table>
```

Let's go through each attribute and its purpose:

- `[value]="products"`: This binds the value of the table to a variable called `products`.
- `[paginator]="true"`: This enables pagination for the table.
- `[rowsPerPageOptions]="[5,10,20]"`: This defines the options for the number of rows to display per page. In this case, the options are set to `5`, `10`, and `20`. The user can select one of these options to control the number of rows displayed in the table.
- `[rows]="10"`: This sets the number of rows to display per page. In this case, it is set to `10`. This means that each page of the table will display up to 10 rows.

Let's take a look at the table with the paginator feature enabled:

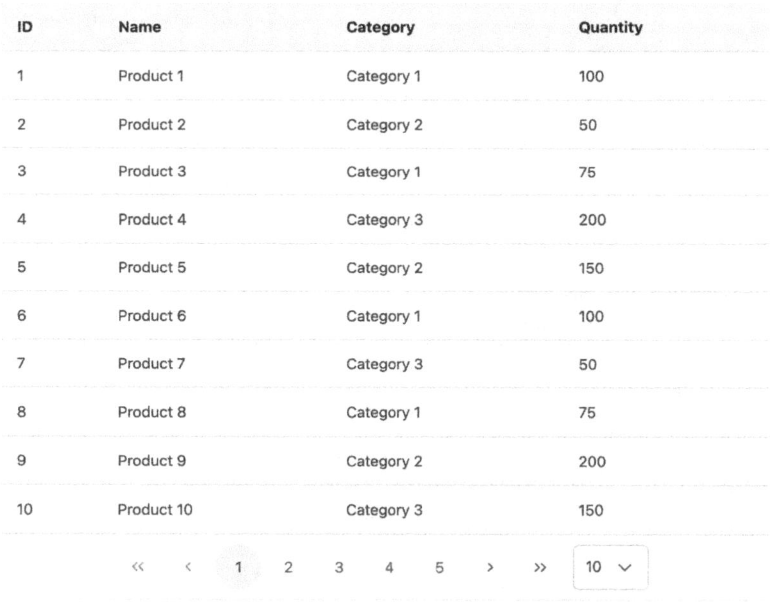

ID	Name	Category	Quantity
1	Product 1	Category 1	100
2	Product 2	Category 2	50
3	Product 3	Category 1	75
4	Product 4	Category 3	200
5	Product 5	Category 2	150
6	Product 6	Category 1	100
7	Product 7	Category 3	50
8	Product 8	Category 1	75
9	Product 9	Category 2	200
10	Product 10	Category 3	150

« ‹ 1 2 3 4 5 › » 10 ⌄

Figure 6.9 – Table with paginator

This screenshot showcases the availability of an option to choose the desired number of rows to be displayed, with a default value set to `10`. Additionally, the paginator feature allows seamless navigation between different pages.

In summary, PrimeNG's table component offers a robust and flexible solution for displaying tabular data. With features such as sorting, filtering, and pagination, it provides developers with the tools needed to present data in an organized and interactive manner.

> **Note**
>
> To learn more about the table component and explore additional features, such as scalable columns, frozen columns, and scrollable tables, you can visit the PrimeNG documentation at `https://primeng.org/table`.

Next, we'll explore another essential aspect of data presentation in web applications: PrimeNG's list components, which offer a diverse set of tools for displaying and interacting with lists of data.

Working with list components

PrimeNG offers a variety of list components that cater to different needs and use cases. These components are designed to transform raw data into meaningful lists, providing users with an intuitive way to interact with the information.

PrimeNG's list components include several key elements that can be used to create diverse list presentations:

- `DataView`: This element offers grid and list views for displaying data, with sorting and filtering options
- `OrderList`: This element allows users to reorder items within a list
- `PickList`: This element enables users to pick items from one list and move them to another

These components are not just about displaying data; they also provide features such as sorting, filtering, and selection, enhancing the user's ability to interact with the data. Let's take a look at each of them in more detail.

DataView

`DataView` is a versatile component that is particularly useful when you need to present a large amount of data in a structured way. It offers various features, such as pagination, sorting, and customizable templates, making it an excellent choice for building data-driven applications.

Here are some scenarios where you might consider using the PrimeNG `DataView` component:

- **Product listings**: If you have an e-commerce application and want to display a list of products with their details, such as name, image, price, and inventory status, the `DataView` component can help you achieve this efficiently
- **Search results**: When users perform a search on your application and you need to present the search results in a clear and organized manner, the `DataView` component can be used to display the results in a grid or list format with pagination and sorting options

- **Dashboard widgets**: When building a dashboard application with multiple widgets displaying different types of data, `DataView` can be used to present the data in each widget consistently, providing a cohesive user experience

In the following subsections, let's consider an example where we have a collection of products that we want to display using the `DataView` component. Each product has properties such as name, category, and price. We'll showcase the products in a list layout, allowing users to browse through them and add items to their shopping cart.

Creating a basic data view

Prior to utilizing the `DataView` component, it is important to ensure that PrimeFlex is installed within your application. This is necessary because `DataView` relies on the `Grid` functionality provided by PrimeFlex to effectively organize and present data. For detailed instructions on installing PrimeFlex, please refer to the previous section within this chapter.

To get started, we need to import the necessary modules from the PrimeNG library:

```
import { DataViewModule } from 'primeng/dataview'
```

Once we have the dependencies installed and imported, we can use the `DataView` component in our Angular template. Here's an example of how we can display the products in a list layout:

```
<p-dataView [value]="products">
   <ng-template pTemplate="list" let-products>
      <div class="col-12" *ngFor="let product of products">
         <div class="flex flex-column xl:flex-row xl:align-items-start
p-4 gap-4">
            <div class="flex flex-column sm:flex-row justify-content-
between align-items-center xl:align-items-start flex-1 gap-4">
                <div class="flex flex-column align-items-center
sm:align-items-start gap-3">
                    <div class="text-2xl font-bold text-900">{{
product.name }}</div>
                    <div class="flex align-items-center gap-3">
                        <span class="flex align-items-center
gap-2">
                            <i class="pi pi-tag"></i>
                            <span class="font-semibold">{{
product.category }}</span>
                        </span>
                    </div>
                </div>
                <div class="flex sm:flex-column align-items-center
sm:align-items-end gap-3 sm:gap-2">
                    <span class="text-2xl font-semibold">{{'$'+
product.price }}</span>
```

```
                        <button pButton icon="pi pi-shopping-cart"
class="md:align-self-end mb-2 p-button-rounded" ></button>
                    </div>
                </div>
                </div>
            </div>
      </ng-template>
</p-dataView>
```

Here is the code breakdown:

- `<p-dataView>`: This is the Angular component from the PrimeNG library used to display data in a view.
- `[value]="products"`: This attribute binding is binding the `products` property to the `value` attribute of the `<p-dataView>` component. It means that the `products` variable in the component's code is the data source for the data view.
- `<ng-template pTemplate="list" let-products>`: This template renders each item in the data view. The `pTemplate` attribute with a value of `"list"` indicates that this template is for the list items. The `let-products` attribute declares a local variable named `products` that represents the `products` array in the data view.

As a result, we created a product list with three items:

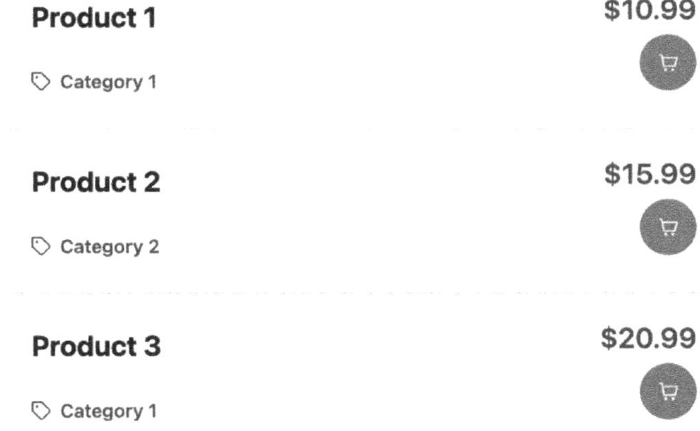

Figure 6.10 – Basic data view

DataView with pagination

If you have a large number of products and want to display them on multiple pages, you can enable pagination in the `DataView` component. Here's an example of how to enable pagination with a specific number of rows per page:

```
<p-dataView
      [value]="products"
      [rows]="4"
      [paginator]="true"
>
      <!-- DataView template -->
</p-dataView>
```

Here, we set the `rows` property to `4`, indicating that we want to display four products per page. By setting the `paginator` property to `true`, the `DataView` component automatically adds pagination controls, allowing users to navigate through the pages.

DataView with sorting

The `DataView` component also provides built-in sorting functionality, allowing users to sort the data based on specific criteria. Here's an example of how to enable sorting and add a dropdown to select the sorting option:

```
<p-dataView
    [value]="products"
    [rows]="4"
    [paginator]="true"
    [sortField]="sortField"
    [sortOrder]="sortOrder"
    >
    <ng-template pTemplate="header">
        <div class="flex flex-column md:flex-row md:justify-content-
between">
            <p-dropdown
                [options]="sortOptions"
                [(ngModel)]="sortKey"
                placeholder="Sort By Price"
                (onChange)="onSortChange($event)"
                styleClass="mb-2 md:mb-0"
            />
        </div>
    </ng-template>
    <!-- DataView template -->
</p-dataView>
```

```
...

sortOptions = [
   { label: 'Price High to Low', value: '!price' },
   { label: 'Price Low to High', value: 'price' }
];

sortOrder!: number;
sortField!: string;

onSortChange(event: HTMLInputElement) {
   const value = event.value;

   if (value.indexOf('!') === 0) {
      this.sortOrder = -1;
      this.sortField = value.substring(1, value.length);
   } else {
      this.sortOrder = 1;
      this.sortField = value;
   }
}
```

Let's break down the code to understand its functionality:

- `[value]="products"`: This binds the `products` array from the parent component to the `DataView` component, which will be the data source for the items displayed.
- `[rows]="4"`: This sets the number of rows to display per page if pagination is enabled.
- `[paginator]="true"`: This enables pagination for the `DataView` component.
- `[sortField]="sortField"` and `[sortOrder]="sortOrder"`: These attributes are used to control the sorting of the data; `sortField` specifies the field by which the data should be sorted, and `sortOrder` specifies the order (ascending or descending).
- `<p-dropdown ... />`: This line creates a dropdown with options defined in the `sortOptions` array. When the user selects an option, the `onSortChange` method is called.
- `sortOptions`: This array defines the sorting options available in the dropdown. The `value` field contains a string that represents the sorting criteria. If the value starts with a `!`, it indicates descending order.
- `onSortChange(event: HTMLInputElement)`: This method is called when the user selects a sorting option from the dropdown. It parses the selected value and sets the `sortOrder` and `sortField` properties accordingly.

By using these properties and templates, you can enable sorting in the `DataView` component and provide a seamless sorting experience to users. In the following screenshot, you can see that we built a product list showing four items at a time and sorted by **Price Low to High**:

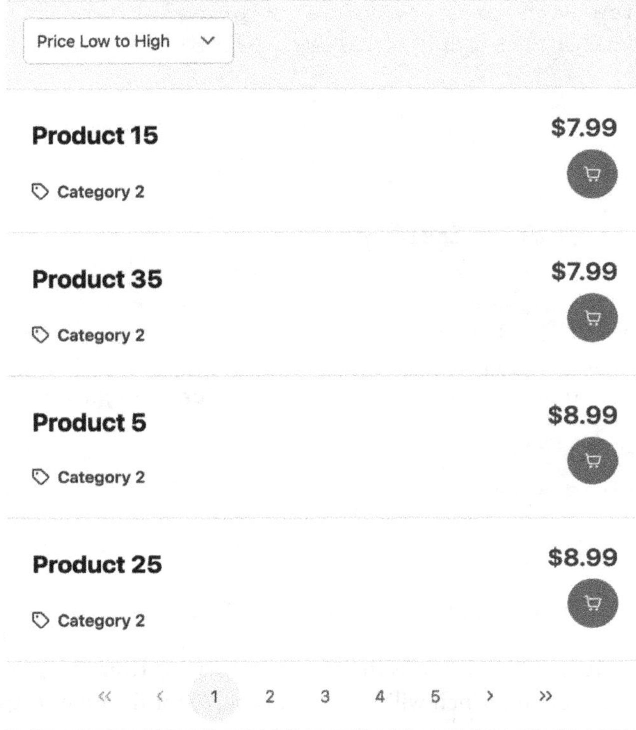

Figure 6.11 – Data view with sorting and pagination

So far, we have witnessed the impressive capabilities of PrimeNG's `DataView` in providing adaptable data presentations. Now, let's explore PrimeNG's `OrderList`, a specialized component that brings a unique touch to list management.

OrderList

When interacting with a collection of items, there are times when the order of these items matters. This is where PrimeNG's `OrderList` comes into play. The `OrderList` component is a powerful tool that allows you to manage and sort a collection of items in your Angular applications. It's like having a personal assistant to help you organize your data.

`OrderList` is especially useful when you need to provide a way for users to manually sort a list of items, such as prioritizing tasks in a to-do list, rearranging chapters in a book, managing playlists, categorizing products, or sorting photos in an album.

Creating a basic order list

Let's dive into an example to see how the `OrderList` component can be used in an application. We'll create a list of products that can be reordered by the user.

The `OrderList` component utilizes `DragDropModule` from the Angular CDK to handle drag and drop actions. It is important to ensure that the `@angular/cdk` package is installed. If not, we need to add it to our `package.json` file by using the following commands:

```
npm install @angular/cdk
```

> **Note**
>
> The `@angular/cdk` (Component Dev Kit or CDK) library is provided by the Angular team and offers a collection of reusable components, directives, and utility functions to simplify the development of Angular applications. The CDK provides a set of tools and building blocks that assist in creating consistent, accessible, and responsive user interfaces. You can learn more about the CDK at `https://material.angular.io/cdk`.

Then, we also need to import `OrderListModule` from PrimeNG to our component:

```
import { OrderListModule } from 'primeng/orderlist'
```

Once we have the dependencies installed and imported, we can use the `OrderList` component in our Angular template. Here's an example of how we can display the products in an order list:

```
<p-orderList
        [value]="products"
        [listStyle]="{ 'max-height': '30rem' }"
        header="Products"
>
   <ng-template let-product pTemplate="item">
      <div class="flex flex-column xl:flex-row xl:align-items-start
gap-4">
         <div
            class="flex flex-column sm:flex-row justify-content-
between align-items-center xl:align-items-start flex-1 gap-4"
         >
            <div
               class="flex flex-column align-items-center sm:align-
items-start gap-3"
            >
               <div class="text-xl font-bold text-700">{{ product.name
}}</div>
               <div class="flex align-items-center gap-3">
                  <span class="flex align-items-center gap-2">
```

```
                    <i class="pi pi-tag"></i>
                    <span class="font-semibold">{{ product.category
}}</span>
                </span>
            </div>
        </div>
        <div
            class="flex sm:flex-column align-items-center sm:align-
items-end gap-3 sm:gap-2"
        >
            <span class="text-xl font-semibold">{{
                '$' + product.price
            }}</span>
            <button
                pButton
                icon="pi pi-shopping-cart"
                class="md:align-self-end mb-2 p-button-rounded"
            ></button>
        </div>
    </div>
  </div>
 </ng-template>
</p-orderList>
```

Here is the breakdown of the example code:

- `[value]="products"`: This binds our list of products to the `value` property of the `OrderList`.
- `header="Products"`: This is used to set a title for the list.
- `[listStyle]="{ 'max-height': '30rem' }"`: This sets the maximum height of the list.
- `<ng-template let-product pTemplate="item">`: This customizes how each item in the list is displayed. We can access the current product using the `let-product` syntax.

Now, in the following screenshot, you'll notice that the product list is labeled as **Products**. Additionally, you have the ability to select a product and reposition it within the list using the arrow buttons located in the left panel:

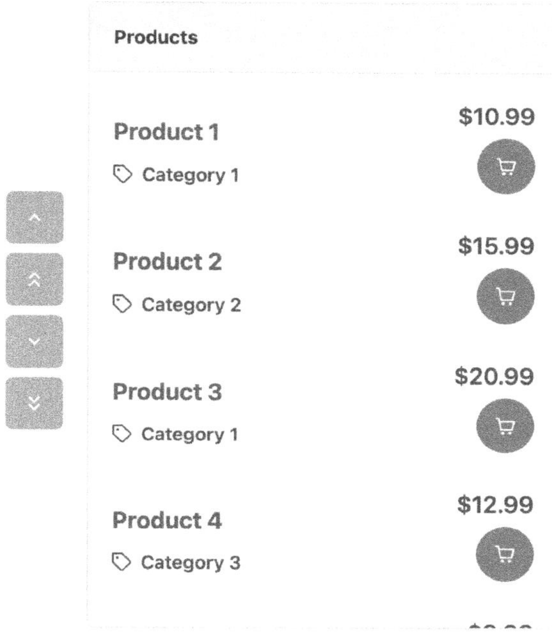

Figure 6.12 – Basic order list

OrderList with filtering

The `OrderList` component also supports filtering and search functionality, allowing users to quickly find specific items within the source list. To enable filtering, we can use the `filterBy` and `filterPlaceholder` properties:

```
<p-orderList
    [value]="products"
    [listStyle]="{ 'max-height': '30rem' }"
    header="Products"
    filterBy="name"
    filterPlaceholder="Filter by name"
>
    <!-- Item template here... -->
</p-orderList>
```

In this example, we set the `filterBy` property to `name` to filter the products based on the product name, and the `filterPlaceholder` property specifies the placeholder text for the search input field. Here is the final result:

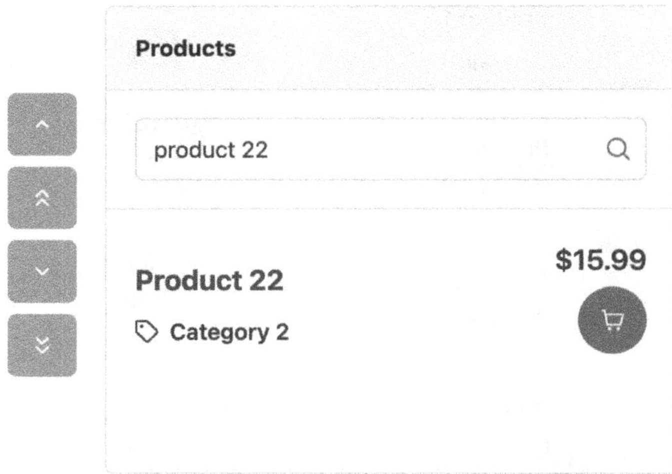

Figure 6.13 – Order list with filtering

As seen in the screenshot, we can filter and search for **Product 22** by typing the name or the product in the search box.

OrderList with drag and drop

The `OrderList` component allows users to reorder items using drag and drop gestures. By default, the drag and drop feature is disabled. To enable it, we can use the `[dragdrop]` property:

```
<p-orderList
    [value]="products"
    [listStyle]="{ 'max-height': '30rem' }"
    header="Products"
    [dragdrop]="true"
>
<!-- Item template here... -->
</p-orderList>
```

Enabling drag and drop allows users to click and hold on to an item, drag it to a new position in the target list, and drop it to reorder the items. This feature provides a visually interactive way for users to rearrange items according to their preferences.

We've discovered the dynamic capabilities of PrimeNG's `OrderList`, enhancing list interactions in our applications. Now, let's shift our focus to PrimeNG's `PickList`, a dual-list interface that promises even more interactivity and versatility.

PickList

The PrimeNG `PickList` component is a powerful tool that allows developers to create interactive and customizable lists for reordering items between different lists. It provides a user-friendly interface for managing and manipulating data in a drag and drop manner. Whether you need to implement a multi-select feature, build a task management system, or create a custom form builder, the `PickList` component offers the flexibility and functionality to meet your requirements.

This can be beneficial in the following scenarios:

- **Task management**: Imagine you are developing a project management application that allows users to assign tasks to different team members. The `PickList` component can be used to display a list of available tasks in the source list and a list of assigned tasks in the target list. Users can easily move tasks between the lists based on their assignment status.
- **Form building**: When building complex forms with a large number of fields, the `PickList` component can assist in organizing the form elements. You can display all available fields in the source list and move selected fields to the target list to define the form structure. This provides a convenient way to dynamically generate forms based on user preferences.
- **Product catalogs**: In an e-commerce application, you may want to provide users with the ability to create custom product catalogs by selecting items from a master list. The `PickList` component can facilitate this process by allowing users to move selected products to the target list, which represents their customized catalog.

By leveraging the `PickList` component, you can enhance the user experience, improve data organization, and enable efficient data manipulation within your application.

Let's consider an example where we have a list of products, and we want to allow users to add selected products to their cart using the PrimeNG `PickList` component. To get started, we need to import the necessary modules from the PrimeNG library:

```
import { PickListModule } from 'primeng/picklist'
```

Once we have the dependencies installed and imported, we can use the `PickList` component in our Angular template. Here's an example of how we can display the products in a pick-list layout:

```
<p-pickList
    [source]="products"
    [target]="selectedProducts"
    sourceHeader="Available Products"
    targetHeader="Selected Products"
```

```
    [dragdrop]="true"
    [responsive]="true"
    [sourceStyle]="{ height: '30rem' }"
    [targetStyle]="{ height: '30rem' }"
    breakpoint="1400px"
>
    <ng-template let-product pTemplate="item">
        <!-- Item template here... -->
    </ng-template>
</p-pickList>
```

In this example, we pass the `products` and `selectedProducts` arrays to the `source` and `target` properties of the `PickList` component, respectively. We also provide labels for the source and target lists using the `sourceHeader` and `targetHeader` properties.

Let's take a look at the final result:

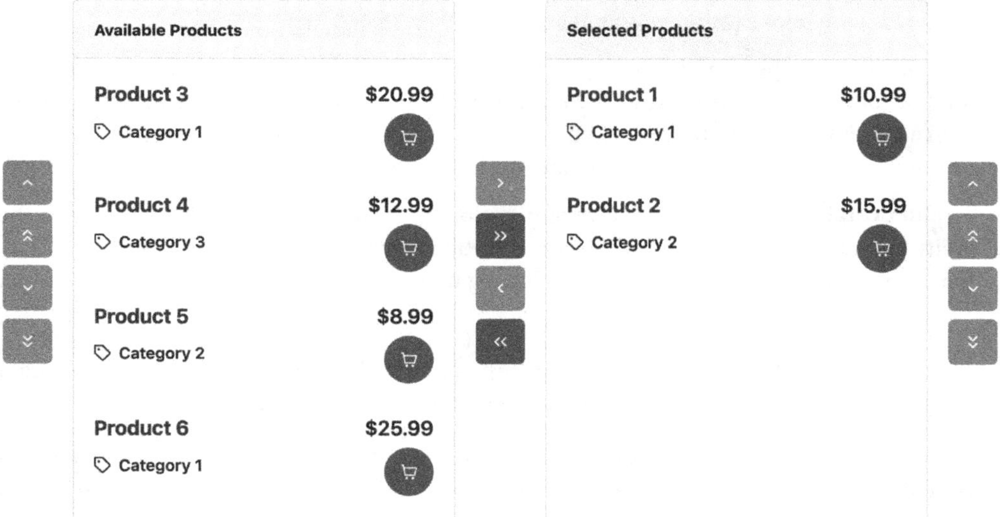

Figure 6.14 – Sample pick list

When you run the application, you should see two lists rendered side by side. The left list represents the available products, and the right list represents the selected products. Users can select products from the source list and move them to the target list by using the drag and drop functionality provided by the `PickList` component.

In summary, PrimeNG's list components are a powerful set of tools for displaying and interacting with lists of data in Angular applications. With their flexibility, customization options, and integration with Angular, they provide developers with everything needed to create engaging and functional list presentations. Whether you need simple lists or more complex interactions such as reordering and selection, PrimeNG's list components offer a robust solution that can enhance any application's user interface.

In the next section, we will go through the final component of this chapter: the PrimeNG Card component.

Working with card components

A PrimeNG card is a container component that provides a flexible and extensible content container with multiple variants and options. It's essentially a rectangular box that holds content and actions about a single subject. Think of it as a small container that groups together specific information.

Cards are incredibly versatile and can be used in various scenarios:

- **Product listings**: Cards can be used for displaying products in an online store where each card represents a product with an image, title, price, and description
- **User profiles**: On social media platforms, cards can represent user profiles, showcasing an image, name, and other personal details
- **Blog posts**: For blog listings, each card might display a post's featured image, title, and a brief summary

The use of cards can make content more digestible, breaking information into chunks that are easier to understand at a glance.

Let's dive into a practical example. Imagine you're building an online store and want to display a list of products using the PrimeNG `Card` component. To get started, we need to import the necessary modules from the PrimeNG library:

```
import { CardModule } from 'primeng/card'
```

After that, we add the PrimeNG `Card` component for creating a product list:

```
<p-card
      ngFor="let product of products"
    [header]="product.name"
    [style]="{ width: '300px' }"
>
    <img
       src="assets/placeholder.png"
       alt="{{ product.name }}"
       style="width:100%"
    />
    <div class="flex flex-column">
       <p>{{ product.description }}</p>
```

```
        <h3>\${{ product.price }}</h3>
        <button pButton type="button" label="Add to Cart"></button>
    </div>
</p-card>
```

This code demonstrates the usage of the PrimeNG `Card` component in an Angular template. Let's break down the code and explain each part:

- `<p-card>`: This is the start of the PrimeNG `Card` component, which represents a single card element
- `*ngFor="let product of products"`: This is an Angular structural directive called `ngFor`, used to iterate over an array of products and generate a card for each product
- `[header]="product.name"`: This binds the `product.name` property to the header input of the `Card` component, which sets the header text of the card
- `[style]="{ width: '300px' }"`: This binds an inline CSS style to the style input of the `Card` component, setting the width of the card to 300 pixels

Let's take a look at the final result:

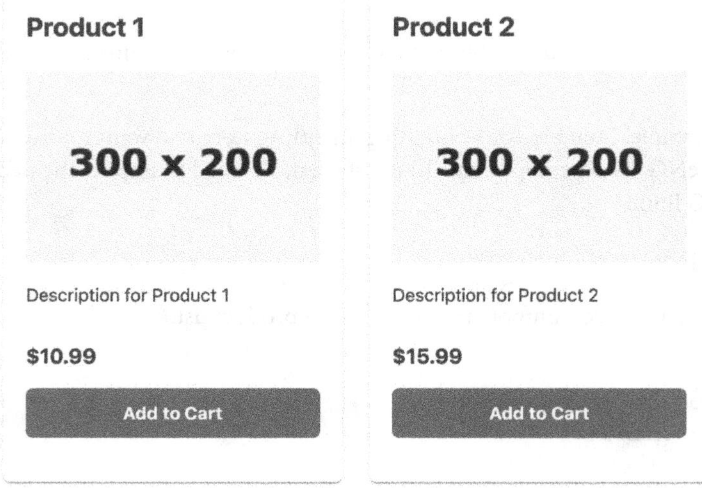

Figure 6.15 – Card example

In this example, we're iterating over a list of products and creating a card for each one. Each card displays the product's image, description, price, and an **Add to Cart** button.

We have recently learned about the flexibility and design features of PrimeNG's card component, which has greatly improved our ability to create visually appealing user interfaces. Now, it's important for us to review and solidify our understanding.

Summary

In this chapter, we've delved into data display components in PrimeNG. We started by understanding the role of these components in transforming raw data into meaningful, user-friendly information, and saw how PrimeNG's rich set of data display components, such as tables, lists, and cards, can be used to present data effectively in Angular applications.

The knowledge gained in this chapter is crucial because effective data presentation is a key aspect of building user-friendly applications. By using PrimeNG's data display components, we can create applications that not only look good but also provide a seamless user experience.

But our journey doesn't end here. Looking ahead to the next chapter, we'll dive into other PrimeNG data display components. We'll learn how to use components such as `Tree`, `Scroller`, `Timeline`, and `VirtualScroller` to show data in our Angular applications.

So, let's keep the momentum going. We've made great strides in understanding and using PrimeNG's data display components. Now, it's time to take the next step and explore data manipulation components. Onward to the next chapter!

7

Working with Tree, TreeTable, and Timeline Components

Data presentation in applications isn't just about tables, lists, and cards. Sometimes, the nature of the data demands a more hierarchical or chronological structure. That's where components such as `Tree`, `TreeTable`, and `Timeline` come into play. In this chapter, we'll dive deep into these specialized components, each offering unique ways to display and interact with data in Angular applications using PrimeNG.

The primary objective is to familiarize yourself with the specialized PrimeNG components that cater to specific data presentation needs. As we progress through this chapter, we'll be equipped with the knowledge to handle diverse data presentation challenges and how to implement them effectively in their Angular applications. Mastering these components means being prepared to offer solutions that enhance user experience and data clarity.

In this chapter, we will cover the following topics:

- Working with Tree components
- Working with TreeTable components
- Working with Timeline components

Technical requirements

This chapter contains various code samples of PrimeNG displaying components. You can find the related source code in the `chapter-07` folder of the following GitHub repository: `https://github.com/PacktPublishing/Next-Level-UI-Development-with-PrimeNG/tree/main/apps/chapter-07`.

Working with Tree components

The PrimeNG **Tree** component is a powerful tool for displaying hierarchical data in an organized and visually appealing way, providing a tree-like structure where data can be presented as nodes and can be expanded or collapsed to reveal or hide child nodes.

In this section, we will explore the various aspects of working with the PrimeNG `Tree` component, including its purpose, usage, and key features. The component also offers a wide range of features, including node expansion and collapse, selection modes, lazy loading of data, drag-and-drop functionality, and context menu support, which we will explore too.

When to use the PrimeNG Tree component

The PrimeNG `Tree` component is particularly useful in scenarios where data needs to be organized and presented in a hierarchical manner. It is commonly used in applications that deal with categories, file directories, organizational structures, and any other data that exhibits a parent-child relationship.

For example, let's consider a product catalog application. The catalog may have categories, subcategories, and products organized in a hierarchical structure. In such a case, the PrimeNG `Tree` component can be used to visually represent the product catalog, allowing users to navigate through the categories and subcategories and select specific products.

Creating a basic Tree component

To better understand how the PrimeNG `Tree` component can be used, let's take a look at an example of a product catalog – mentioned in the previous section. Suppose we have the following hierarchical structure for our products:

```
- Electronics
   - Computers
      - MacBook Air
      - Smartphone
   - Phones
      - iPhone
      - Samsung
      - Google Pixel
- Home & Garden
   - Outdoor
   - Furniture
   - Office
- Books & Media
   - Books
   - Movies & TV
```

Using the PrimeNG `Tree` component, we can represent this hierarchical structure.

To get started, we need to import the necessary modules from the PrimeNG library:

```
import { TreeModule } from 'primeng/tree'
```

Once we have the dependencies installed and imported, we can use the `Tree` component in our Angular template. Here's an example of how we can display the products in a tree layout:

```
// tree.component.ts
import { TreeNode } from 'primeng/api'

// Html / template
<p-tree [value]="products" />

// TypeScript
products: TreeNode[] = [
    {
        "key": "0",
        "label": "Electronics",
        "data": "Category Level",
        "icon": "pi pi-tablet",
        "children": [
            {
                "key": "0-0",
                "label": "Computers",
                "data": "SubCategory Level",
                "icon": "pi pi-desktop",
                "children": [
                    {
                        "key": "0-0-0",
                        "label": "MacBook Air",
                        "data": "Product Level",
                        "icon": "pi pi-apple"
                    },
                    ...

                ]
            },
            ...
        ]
    }
    ...
]
```

Let's break down the code:

- `<p-tree [value]="products" />`: This represents the usage of the PrimeNG `Tree` component. It binds the `value` property of the `Tree` component to the `products` variable.
- `Products: TreeNode[]`: This defines the `products` variable as an array of `TreeNode` objects. `TreeNode` is a type defined by PrimeNG for representing a node in the tree component. Each `TreeNode` object has the following properties:
  - `key`: A unique identifier for the node.
  - `label`: The text that will be displayed for the node.
  - `data`: Additional data associated with the node. In this example, it represents the level of the node (category level, subcategory level, or product level).
  - `icon`: An optional icon associated with the node.
  - `children`: An array of child nodes. This property allows the tree to have a nested structure.

That is an example of a simplified tree structure representing a store. Here is the result:

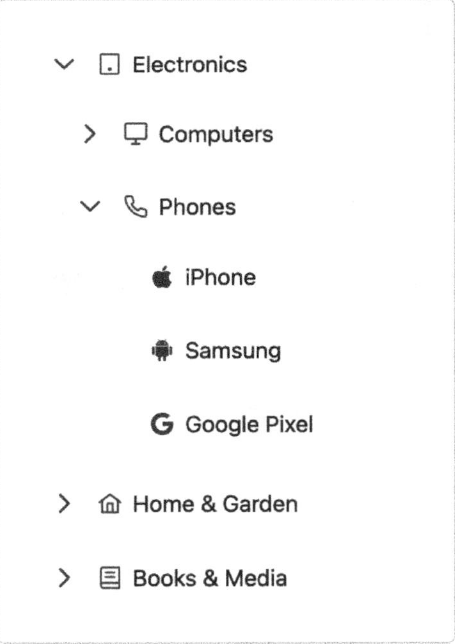

Figure 7.1 – Basic Tree

Expanding and collapsing nodes

One of the fundamental features of the PrimeNG `Tree` component is the ability to expand and collapse nodes. This allows users to navigate through the tree and reveal or hide child nodes based on their interests.

By default, the PrimeNG `Tree` component starts with all nodes collapsed. Users can expand a node by clicking on its expand icon, and they can collapse a node by clicking on its collapse icon.

In addition to user interactions, you can also programmatically control the expansion and collapse of nodes by manipulating the state of the nodes. For example, you can programmatically expand all nodes or collapse all nodes using the `expandAll()` and `collapseAll()` methods, respectively.

Here's an example that demonstrates the use of the expand and collapse functionality:

```
// tree.component.ts
<div class="grid gap-2 p-2 mb-2">
   <button
      pButton
      type="button"
      label="Expand all"
      (click)="expandAll()"
   ></button>
   <button
      pButton
      type="button"
      label="Collapse all"
      (click)="collapseAll()"
   ></button>
</div>

<p-tree [value]="products" />

...

expandAll() {
   this.products.forEach((node) => {
      this.expandRecursive(node, true)
   })
}

collapseAll() {
   this.products.forEach((node) => {
      this.expandRecursive(node, false)
   })
```

```
}

private expandRecursive(node: TreeNode, isExpand: boolean) {
    node.expanded = isExpand
    if (node.children) {
        node.children.forEach((childNode) => {
            this.expandRecursive(childNode, isExpand)
        })
    }
}
```

Here, we have added two buttons to the UI for expanding and collapsing nodes. The `expandAll()` and `collapseAll()` methods are bound to the respective buttons' click events. When a user clicks the **Expand all** button, all nodes in the tree will be expanded, and when the **Collapse all** button is clicked, all nodes will collapse.

The `expandRecursive(...)` method is a private method that recursively expands or collapses the nodes in the tree. It takes a `TreeNode` object (`node`) and a `boolean` value (`isExpand`) as parameters, and sets the `expanded` property of `node` to the value of `isExpand`, thus expanding or collapsing the node.

Here is the result when clicking on the **Expand all** button:

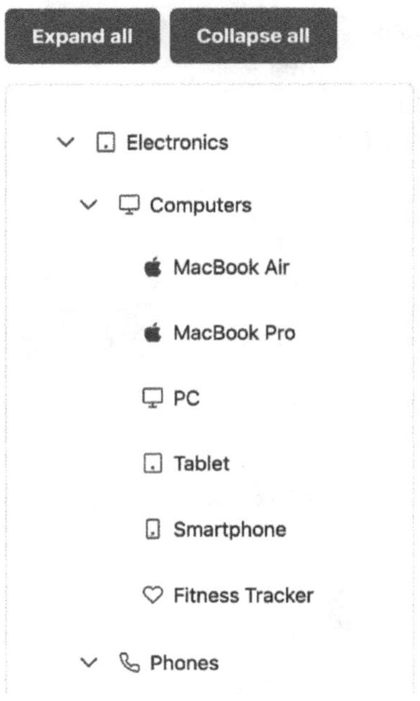

Figure 7.2 – Tree with expanding and collapsing abilities

Using node selection events

The PrimeNG `Tree` component provides a set of events and methods that allow you to interact with the component and respond to user actions. These events and methods enable you to perform tasks such as handling node selection, capturing node expansion and collapse events, and dynamically loading data.

Before enabling node selection events, we need to add `selectionMode` to the component. There are four types of selection:

- **Single selection**: As the name suggests, this mode allows users to select only one node at a time. It's ideal for scenarios where a single choice is required from the user – for example, `<p-tree [value]="products" selectionMode="single" />`.
- **Multiple selection with metaKey**: In this mode, users can select multiple nodes, but they need to hold down a metakey (*Ctrl* or *Shift*) while making their selections – for example, `<p-tree [value]="products" selectionMode="single" [metaKeySelection]="true"/>`.
- **Multiple selection without metaKey**: This mode allows users to freely select multiple nodes without the need for any additional key presses – for example, `<p-tree [value]="products" selectionMode="multiple" />`.
- **Checkbox selection**: This mode provides checkboxes next to each node, allowing users to make multiple selections more intuitively – for example, `<p-tree [value]="products" selectionMode="checkbox" />`.

After adding the node selection type, the PrimeNG `Tree` component emits events when a node is selected or unselected. You can use these events to perform actions based on the user's node selection.

To capture the node selection event, you can use the `(onNodeSelect)` and `(onNodeUnselect)` event bindings. Here's an example of `single` selection:

```
// tree.component.ts
<p-tree
      [value]="products"
      selectionMode="single"
      (onNodeSelect)="onNodeSelected($event)"
      (onNodeUnselect)="onNodeUnselected($event)"
/>

...

onNodeSelected(event: TreeNodeSelectEvent) {
   console.log(event)
}
```

```
onNodeUnselected(event: TreeNodeSelectEvent) {
   console.log(event)
}
```

In the code, the `(onNodeSelect)` event is bound to the `onNodeSelected()` method, and the `(onNodeUnselect)` event is bound to the `onNodeUnselected()` method. These methods will be invoked when a node is selected or unselected, respectively.

Let's have a look at a sample event when selecting the **Electronics** node:

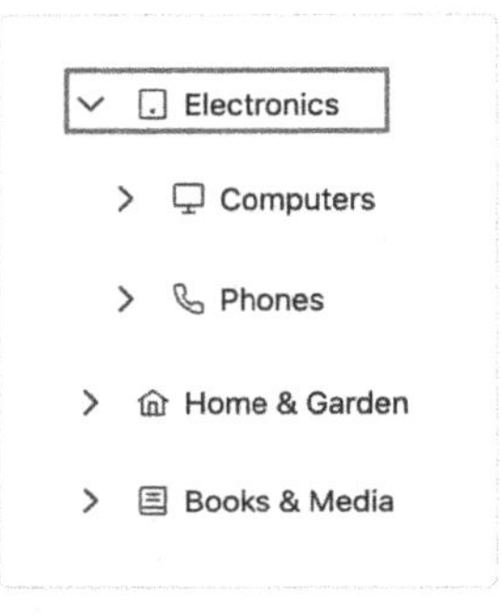


```
                                                                  tree.component.ts:72
▾ {originalEvent: PointerEvent, node: {…}}
  ▾ node:
    ▾ children: Array(2)
      ▾ 0:
        ▸ children: (6) [{…}, {…}, {…}, {…}, {…}, {…}]
          data: "SubCategory Level"
          expanded: true
          icon: "pi pi-desktop"
          key: "0-0"
          label: "Computers"
        ▸ parent: {key: '0', label: 'Electronics', data: 'Category Level', icon: 'pi pi-tablet'
        ▸ [[Prototype]]: Object
      ▾ 1:
        ▸ children: (3) [{…}, {…}, {…}]
          data: "SubCategory Level"
          icon: "pi pi-phone"
          key: "0-1"
          label: "Phones"
        ▸ parent: {key: '0', label: 'Electronics', data: 'Category Level', icon: 'pi pi-tablet'
        ▸ [[Prototype]]: Object
        length: 2
      ▸ [[Prototype]]: Array(0)
      data: "Category Level"
      expanded: true
      icon: "pi pi-tablet"
      key: "0"
      label: "Electronics"
      parent: undefined
    ▸ [[Prototype]]: Object
  ▸ originalEvent: PointerEvent {isTrusted: true, pointerId: 1, width: 1, height: 1, pressure:
  ▸ [[Prototype]]: Object
```

Figure 7.3 – Tree with selection events

As a result, we can see that after selecting the **Electronics** node, we gain access to the `node` data, which has the following details:

- `expanded: true`: This indicates the current node is in the expanded state
- `parent: undefined`: This shows that the current node doesn't have a parent
- `data, icon, key, label, children`: This shows the existing values of the current node

Using node expansion and collapse events

The PrimeNG `Tree` component emits events when a node is expanded or collapsed. You can utilize these events to perform actions when nodes are expanded or collapsed by the user.

To capture node expansion or collapse events, you can use the `(onNodeExpand)` and `(onNodeCollapse)` event bindings. Here's an example:

```
// tree.component.ts
<p-tree
    [value]="productsWithEvents"
    (onNodeExpand)="onNodeExpanded($event)"
    (onNodeCollapse)="onNodeCollapsed($event)"
/>

...

onNodeExpanded(event: TreeNodeSelectEvent) {
    console.log(event)
}

onNodeCollapsed(event: TreeNodeSelectEvent) {
    console.log(event)
}
```

Here, the `(onNodeExpand)` event is bound to the `onNodeExpanded()` method, and the `(onNodeCollapse)` event is bound to the `onNodeCollapsed()` method. These methods will be triggered when a node is expanded or collapsed, respectively. The event value after expanding or collapsing is the same as when we select or unselect a node.

Working with lazy loading

The PrimeNG `Tree` component supports lazy loading of data, which is beneficial when dealing with large datasets. Instead of loading all the nodes at once, you can load nodes dynamically as the user expands them.

To enable lazy loading, you need to use the `[loading]` property and the `(onNodeExpand)` event. The `[loading]` property allows you to indicate whether the tree is currently loading data, and the `(onNodeExpand)` event is triggered when a node is expanded, allowing you to load the child nodes dynamically.

Here's an example that demonstrates lazy loading:

```
<p-tree
    [loading]="loading"
    [value]="products"
    (onNodeExpand)="loadChildNodes($event)"
/>

...
```

```
loading = false

loadChildNodes(event: TreeNodeSelectEvent) {
    if (event.node) {
        this.loading = true

        // example of retrieving child nodes data
        event.node.children = this.nodeService.getChildNodes(event.node)
        this.loading = false
    }
}
```

In the preceding code, the `[loading]` property is bound to the `loading` variable in the Angular component, which indicates whether the tree is currently loading data. The `(onNodeExpand)` event is bound to the `loadChildNodes()` method, which is responsible for loading the child nodes of the expanded node. Here is the result:

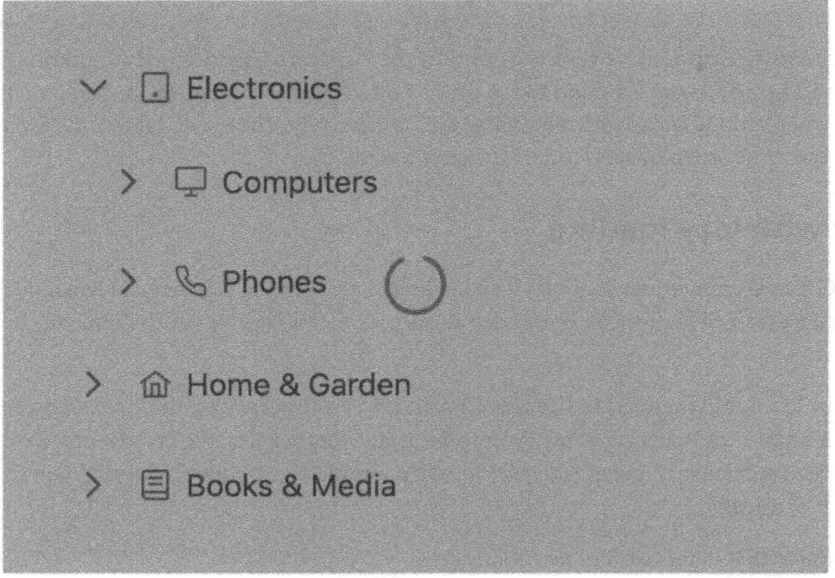

Figure 7.4 – Tree with loading

> **Note**
>
> The `loadChildNodes()` method should be implemented in the Angular component to fetch the child nodes dynamically based on the expanded node.

In this section, we discussed when to use the PrimeNG `Tree` component, such as in scenarios where data needs to be organized hierarchically, such as product catalogs or file systems, and its various features. In the next section, let's delve into the `TreeTable` component.

Working with TreeTable components

When it comes to presenting hierarchical data structures in a tabular format, PrimeNG's **TreeTable** emerges as a powerful tool, combining the best of both worlds: the nested structure of a tree and the organized columns of a table. Let's embark on a journey to understand this component better and see how it can elevate our data presentation game.

When to use the PrimeNG TreeTable component

The PrimeNG `TreeTable` component is particularly useful in scenarios where you need to represent data that has a hierarchical structure. It provides an intuitive and user-friendly way to navigate and interact with hierarchical data, making it ideal for applications that deal with organizational structures, file systems, product categories, and any other data that exhibits parent-child relationships.

The `TreeTable` component is more suitable for presenting hierarchical data in a tabular format with advanced interaction options such as sorting, filtering, and pagination, while the `Tree` component is best suited for displaying and navigating hierarchical data in a compact and collapsible tree-like structure.

By leveraging the `TreeTable` component, you can present complex hierarchical data in a structured and organized manner, allowing users to expand and collapse nodes, perform sorting and filtering operations, and interact with the data in a seamless manner.

Creating a basic TreeTable component

Imagine you have a list of products categorized under different product families. Each product has details such as price, availability, and ratings. The `TreeTable` component can be an excellent choice to represent this data.

To get started, we need to import the necessary modules from the PrimeNG library:

```
import { TreeTableModule } from 'primeng/treetable'
```

Once we have the dependencies installed and imported, we can use the `TreeTable` component in our Angular template. Here's an example of how we can display the products in a `TreeTable` layout:

```
<p-treeTable
    [value]="products"
    [scrollable]="true"
    [tableStyle]="{ 'min-width': '50rem' }"
>
    <ng-template pTemplate="header">
        <tr>
            <th>Name</th>
            <th>Price (USD)</th>
            <th>Rating</th>
        </tr>
    </ng-template>

    <ng-template pTemplate="body" let-rowNode let-product="rowData">
        <tr [ttRow]="rowNode">
            <td>
                <p-treeTableToggler [rowNode]="rowNode" />
                {{ product.name }}
            </td>
            <td>{{ product?.price | currency }}</td>
            <td>{{ product?.rating }}</td>
        </tr>
    </ng-template>
</p-treeTable>

...

products: TreeTableNode[] = [
    {
        "key": "0",
        "data": {
            "name": "Electronics"
        },
        "children": [
            {
                "key": "0-0",
                "data": {
                    "name": "Computers"
                },
                "children": [
                    {
```

```
                "key": "0-0-0",
                "data": {
                  "id": 1,
                  "name": "MacBook Air",
                  "price": 999,
                  "description": "Light and portable MacBook",
                  "quantity": 100,
                  "rating": 4,
                  "category": "Computers"
                }
              },
              ...
            ]
          }
        ]
      }
      ...
    ]
```

Let's break down the code:

- `<p-treeTable>`: This is the Angular component from the PrimeNG library used to display hierarchical tabular data in a tree-like structure.
- `[value]="products"`: This attribute binding sets the `products` property as the data source for the `TreeTable` component. The `products` variable in the component's code contains an array of `TreeTableNode` objects representing the hierarchical data structure.
- `[scrollable]="true"`: This attribute binding enables scrolling within the `TreeTable` component if the content overflows the available space.
- `[tableStyle]="{ 'min-width': '50rem' }"`: This attribute binding applies an inline CSS style object to the `tableStyle` attribute of the `<p-treeTable>` component. In this case, it sets the minimum width of the `TreeTable` component to `50rem`.
- `<ng-template pTemplate="header">`: This element defines a template for rendering the header row of the `TreeTable` component.
- `<ng-template pTemplate="body" let-rowNode let-product="rowData">`: This element defines a template for rendering the body (rows) of the `TreeTable` component. This template also has two other properties:
  - `let-rowNode`: This declares a local variable named `rowNode` that represents the current row node being rendered
  - `let-product="rowData"`: This declares a local variable named `product` that represents the data associated with the current row

- <p-treeTableToggler [rowNode]="rowNode" />: This component is used to display a toggle button for expanding and collapsing child nodes in the TreeTable component.
- products: TreeTableNode[]: This defines the products array as the data source for the TreeTable component. The products array consists of TreeTableNode objects representing the hierarchical data structure. Each node has a data property containing the product information, such as name, price, rating, and category. The example shows a nested structure with a parent node, Electronics, and a child node, Computers, which further contains a MacBook Air child node with its respective properties.

That is an example of a simplified TreeTable structure representing a store. Here is the result:

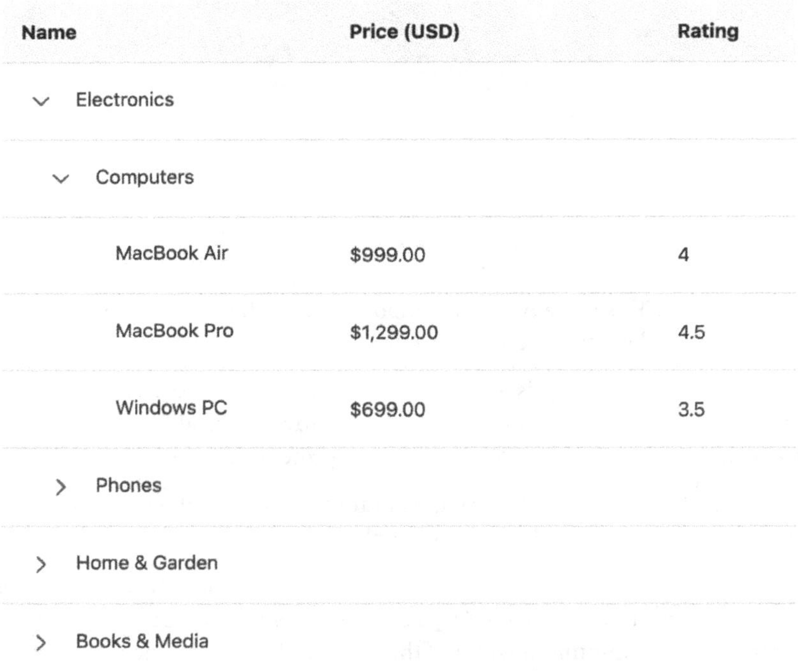

Name	Price (USD)	Rating
∨ Electronics		
∨ Computers		
MacBook Air	$999.00	4
MacBook Pro	$1,299.00	4.5
Windows PC	$699.00	3.5
> Phones		
> Home & Garden		
> Books & Media		

Figure 7.5 – Basic TreeTable structure

Using dynamic columns

Dynamic columns in PrimeNG's TreeTable component allow us to create columns on the fly based on the data or configuration we provide. Instead of defining each column manually in the template, we can bind a collection of columns from our component and let the TreeTable component generate the necessary columns dynamically. This approach is not only efficient but also offers a high degree of flexibility.

Let's consider an example of an e-commerce application that displays products in a `TreeTable` format. The application needs to handle different product categories, each with its own set of attributes. The goal is to dynamically render the columns based on the selected product category. Here is the code:

```
<div class="grid gap-4 ml-0 mb-4">
    <button
        (click)="updateColumns('RATING')"
        pButton
        label="Rating"
    ></button>

    <button
        (click)="updateColumns('QUANTITY')"
        pButton
        label="Quantity"
    ></button>
</div>

<p-treeTable
    [value]="products"
    [columns]="cols"
    [scrollable]="true"
    [tableStyle]="{ 'min-width': '50rem' }"
>
    <ng-template pTemplate="header" let-columns>
        <tr>
            <th *ngFor="let col of columns">
                {{ col.header }}
            </th>
        </tr>
    </ng-template>

    <ng-template
        pTemplate="body"
        let-rowNode
        let-product="rowData"
        let-columns="columns"
    >
        <tr>
            <td *ngFor="let col of columns; let i = index">
                <p-treeTableToggler [rowNode]="rowNode" *ngIf="i === 0" />
                {{ product[col.field] }}
            </td>
        </tr>
```

```
    </ng-template>
</p-treeTable>

...

cols = [
    { field: 'name', header: 'Name' },
    { field: 'price', header: 'Price' },
    { field: 'rating', header: 'Rating' },
]

updateColumns(option: string) {
    switch (option) {
        case 'RATING':
            this.cols = [
                { field: 'name', header: 'Name' },
                { field: 'price', header: 'Price' },
                { field: 'rating', header: 'Rating' },
            ]
            break

        case 'QUANTITY':
            this.cols = [
                { field: 'name', header: 'Name' },
                { field: 'price', header: 'Price' },
                { field: 'quantity', header: 'Quantity' },
            ]
            break

        default:
            break
    }
}
```

Let's break down the code:

- `<button (click)="updateColumns(...)" >`: This is a button element that triggers the `updateColumns()` method when clicked.
- `<th *ngFor="let col of columns">{{ col.header }}</th>`: This line uses an `*ngFor` directive to iterate over the `columns` array and generates a `<th>` element for each column. The column's `header` property is displayed as the content of the `header` cell.
- `<td *ngFor="let col of columns; let i = index">`: This line uses an `*ngFor` directive to iterate over the `columns` array and generates a `<td>` element for each column.

- `<p-treeTableToggler [rowNode]="rowNode" *ngIf="i === 0" />`: This `<p-treeTableToggler>` component is used to display a toggle button for expanding and collapsing child nodes in the `TreeTable` component.

With this implementation, the `TreeTable` will display the product data with the appropriate columns based on the selected type. Let's look at the result:

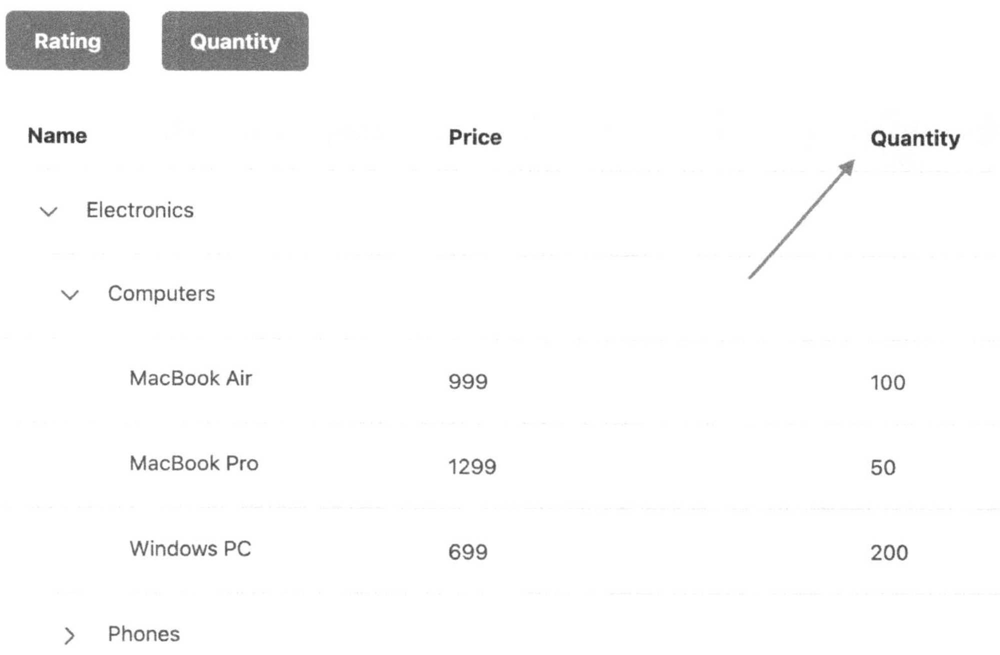

Figure 7.6 – TreeTable with dynamic columns

If the user switches from the **Rating** type to the **Quantity** type, the table will automatically update to show the **Quantity** column instead of **Rating**, as depicted back in *Figure 7.5*.

Enabling the TreeTable paginator

The **paginator** is a feature within PrimeNG's `TreeTable` component that allows us to break down large sets of data into smaller, more manageable chunks or pages. Instead of displaying hundreds or thousands of rows all at once, the paginator lets users navigate through the data one page at a time. It provides controls for moving to the next page or previous page, jumping to the start or the end, and even selecting the page size.

We can easily enable the paginator in our `TreeTable` component by adding `paginator` and `rows` attributes, as shown in the following code:

```
<p-treeTable [value]="products" [paginator]="true" [rows]="2">
    <!-- Column templates and other TreeTable configurations go here
-->
</p-treeTable>
```

In this example, we've enabled pagination by setting `[paginator]="true"` and specified that we want to display two rows per page with `[rows]="2"`.

The `TreeTable` component will now display pagination controls, and users can navigate through the product data page by page:

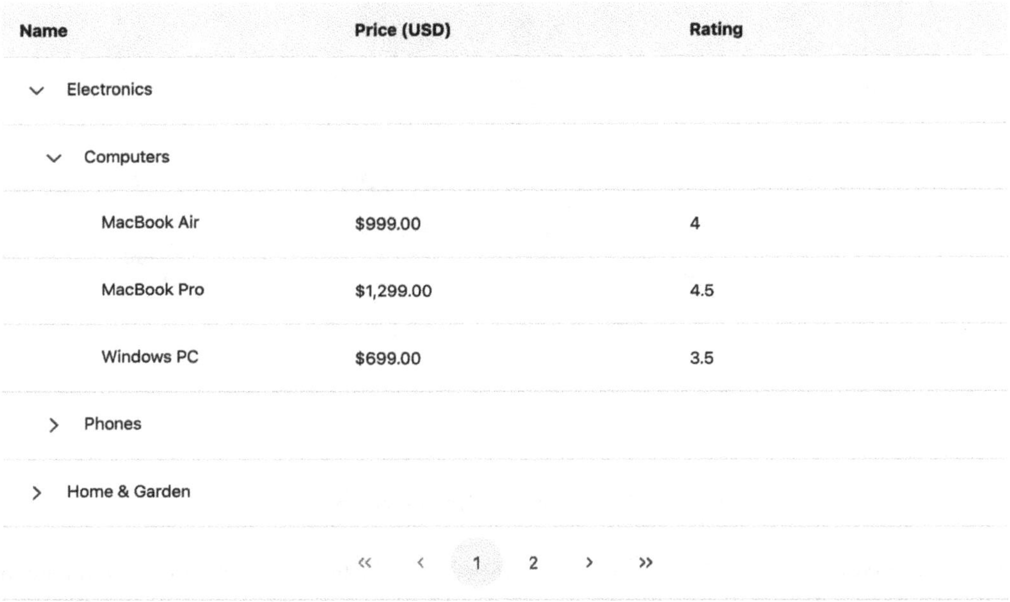

Name	Price (USD)	Rating
∨ Electronics		
∨ Computers		
MacBook Air	$999.00	4
MacBook Pro	$1,299.00	4.5
Windows PC	$699.00	3.5
> Phones		
> Home & Garden		

« ‹ 1 2 › »

Figure 7.7 – TreeTable with paginator

Associated events and methods

The `TreeTable` component provides various events and methods that you can leverage to enhance the functionality and interactivity of your application. Here are some commonly used events and methods:

- **Events**:
    - `onNodeExpand`: This event is triggered when a node is expanded

    - `onNodeCollapse`: This event is triggered when a node is collapsed
    - `onNodeSelect`: This event is triggered when a node is selected
    - `onNodeUnselect`: This event is triggered when a node is unselected

- **Methods**:

    - `reset`: This method clears the sort and paginator state
    - `resetScrollTop`: This method resets scroll to top
    - `scrollToVirtualIndex`: This method scrolls to a given index when using virtual scrolling
    - `scrollTo`: This method scrolls to the given index

You can use these events and methods to customize the behavior of the `TreeTable` component based on your application's requirements. For example, you can listen to the `onNodeCollapse` event to perform certain actions when a node is collapsed, such as removing detailed information or triggering additional operations:

```
<p-treeTable
[value]="files"
(onNodeCollapse)="handleNodeCollapse($event)"
>
      <!-- Column templates and other TreeTable configurations -->
</p-treeTable>

...

handleNodeCollapse(event: TreeTableNodeCollapseEvent) {
   const collapsedNodeData = event.node.data
   // Handle actions when a node is collapsed
}
```

Let's break down the code:

- `(onNodeCollapse)="handleNodeCollapse($event)"`: This binds the `handleNodeCollapse` method to the event that occurs when a node in the `TreeTable` component is collapsed. The method will be called with the `event` object as an argument.
- `handleNodeCollapse(event: TreeTableNodeCollapseEvent) { ... }`: This takes the `event` object of the `TreeTableNodeCollapseEvent` type as an argument. This `event` object contains information about the collapsed node, which you can use to handle the event.

Throughout our exploration of the PrimeNG `TreeTable` component, we've seen its power in presenting hierarchical data in a structured and user-friendly manner. From dynamic columns to efficient pagination, the `TreeTable` component offers a robust solution for various data representation challenges. Now, let's shift our focus to the PrimeNG `Timeline` component, a tool that beautifully visualizes data in a chronological sequence.

Working with Timeline components

The PrimeNG `Timeline` component is a powerful component provided by the PrimeNG library for Angular applications, allowing you to visualize a series of chained events in chronological order. The timeline provides a user-friendly and interactive way to display events, making it easier for users to understand the sequence of activities or changes over time.

The PrimeNG `Timeline` component is designed to present events in a linear fashion, allowing users to navigate through different stages or milestones. Each event in the timeline is represented by a marker, which can be customized to show relevant information such as status, date, or any other meaningful data.

The timeline offers various features to enhance the user experience – it supports both vertical and horizontal layouts, providing flexibility in terms of the orientation of the timeline, plus alignment options to position the timeline bar relative to the content.

When to use the PrimeNG Timeline component

The PrimeNG `Timeline` component can be used in a wide range of applications and scenarios. Here are a few examples of when the `Timeline` component can be beneficial:

- **Project management**: Use a timeline to showcase project milestones, such as project initiation, requirements gathering, development phases, and project completion. It helps stakeholders and team members visualize the project's progress and understand the sequence of key events.
- **Order tracking**: If you have an e-commerce application, you can utilize a timeline to display the different stages of order processing, such as order placement, payment verification, order fulfillment, and delivery. This gives customers a clear overview of the order's progress.
- **Historical events**: A timeline is also suitable for presenting historical events or significant achievements. For instance, you can use it to illustrate the timeline of scientific discoveries, major historical events, or the evolution of a specific industry.
- **Product updates**: If you maintain a product roadmap or want to showcase the release history of your software product, a timeline can be an effective way to present the different versions, updates, and new features over time.

Whenever there's a need to represent a series of events in the order they took place, the PrimeNG `Timeline` component is an excellent choice.

Creating a basic timeline

To get started, we need to import the necessary modules from the PrimeNG library:

```
import { TimelineModule } from 'primeng/timeline'
```

Once we have the dependencies installed and imported, we can use the `Timeline` component in our Angular template. Here's an example of how we can display the order status in a timeline layout:

```
<p-timeline [value]="orderStatuses">
    <ng-template pTemplate="content" let-order>
        <span [class]="order.icon"></span>
        {{ order.title }}
    </ng-template>
</p-timeline>

...

orderStatuses = [
    {
        title: 'Order Placed',
        content: 'Your order has been received and is being processed.',
        icon: 'pi pi-shopping-cart',
    },
    {
        title: 'Order Confirmed',
        content:
            'Your payment has been confirmed and the order is now being
prepared.',
        icon: 'pi pi-check-square',
    },
    {
        title: 'In Warehouse',
        content: 'Your product is in the warehouse, awaiting dispatch.',
        icon: 'pi pi-globe',
    },
    {
        title: 'Shipped',
        content:
            'Your order has been shipped and is on its way to the
delivery address.',
```

```
        icon: 'pi pi-truck',
    },
    {
        title: 'Out for Delivery',
        content: 'The product is out for delivery and will reach you
soon.',
        icon: 'pi pi-map-marker',
    },
    {
        title: 'Delivered',
        content: 'Your product has been successfully delivered. Enjoy!',
        icon: 'pi pi-check-circle',
    },
]
```

Let's break down the code and understand its functionality:

- `<p-timeline [value]="orderStatuses">`: This represents the usage of the PrimeNG `Timeline` component. It binds the `value` property of the `Timeline` component to the `orderStatuses` variable.
- `<ng-template pTemplate="content" let-order>`: This defines the template for rendering the content of each status in the timeline.
- `orderStatuses`: This represents different stages or statuses of an order. Each element in the array corresponds to a specific event that will be displayed on the timeline.

Overall, the code demonstrates how to use the PrimeNG `Timeline` component to display a timeline of order status. The timeline is populated with status data from the `orderStatuses` array, and each status is rendered using a template that includes an icon and a title. This allows for a visually appealing and informative representation of events in chronological order.

Here is the result of the code:

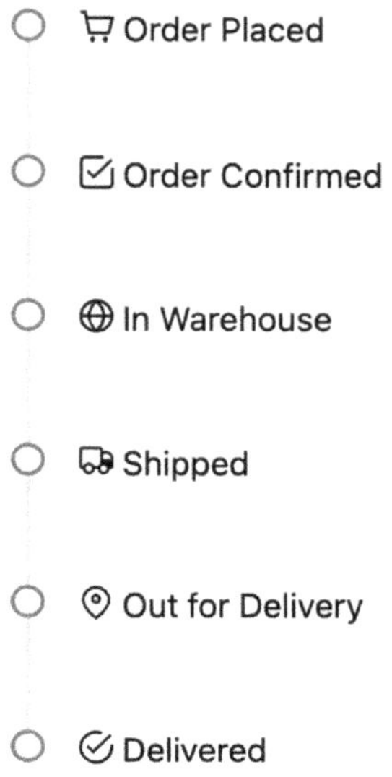

Figure 7.8 – Basic timeline

Timeline alignment

The PrimeNG **timeline alignment** feature enables you to control the location of the content relative to the timeline. You can align the content to the left, right, top, bottom, or alternate sides of the timeline. This flexibility allows you to customize the appearance and layout of the timeline based on your design preferences or application needs.

Let's imagine we're building a timeline to showcase the order statuses of a product. You can customize the alignment by choosing different values for the `align` attribute, such as `left`, `right`, or `alternate`, depending on your specific design requirements. Let's update the alignment in our existing order status timeline:

```
<p-timeline [value]="orderStatuses" align="alternate">
   // timeline content
</p-timeline>
```

In the code snippet, we have used the `align` attribute with the `alternate` value to align the content of each activity on alternate sides of the timeline line. This layout creates an interesting visual pattern, with activities appearing on both the left and right sides of the timeline:

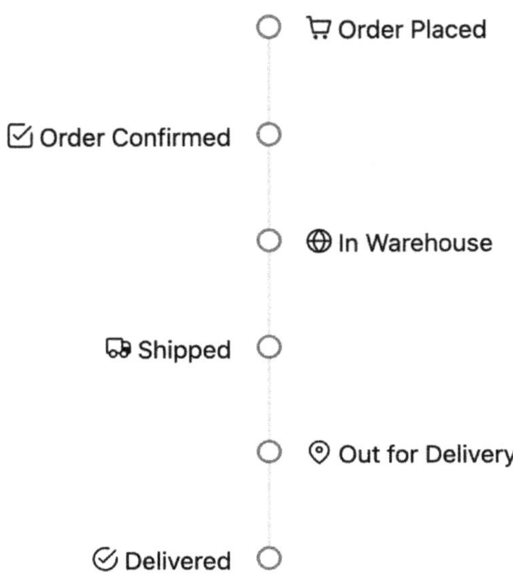

Figure 7.9 – Timeline alignment

Timeline horizontal layout

In addition to the alignment options, the PrimeNG `Timeline` component also provides a horizontal layout option. The horizontal layout is designed to present events or milestones in a linear fashion from left to right, which is particularly useful when you want to showcase a timeline that spans across a wide area, such as a project timeline or a historical sequence of events.

To use `Timeline` with a horizontal layout, you can set the `layout` attribute to `horizontal` in the `Timeline` component. Let's take a look at an example:

```
<p-timeline [value]="orderStatuses" layout="horizontal">
   // timeline content
</p-timeline>
```

In the code snippet, we have set the `layout` attribute to `horizontal`, indicating that we want to display the timeline events in a horizontal manner. Here is the result:

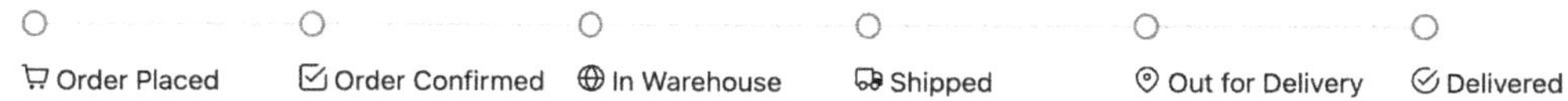

Figure 7.10 – Horizontal timeline

Through our exploration of the PrimeNG `Timeline` component, we've seen how it offers a dynamic way to visually represent events, milestones, or processes. Its flexibility, from basic event representation to features such as alignment or layout, ensures that we can craft narratives tailored to our needs. As we wrap up this section, let's take a moment to reflect on our journey and summarize the key takeaways.

Summary

Navigating through the chapter, we explored the intricacies of representing hierarchical and chronological data using PrimeNG's `Tree`, `TreeTable`, and `Timeline` components. These powerful tools play a pivotal role in presenting structured data in a visually appealing and user-friendly manner, whether it's displaying a hierarchical structure of items or visualizing a sequence of events over time.

We began by uncovering the capabilities of the `Tree` component and understanding its significance in effectively representing data with parent-child relationships. The `TreeTable` component expanded on this concept by offering a seamless integration of tabular and hierarchical data presentation. Additionally, the `Timeline` component showcased its prowess in visualizing sequences, milestones, or events in chronological order, providing us with the flexibility to present narratives or process flows with clarity and coherence.

By mastering these components, we equip ourselves with the necessary tools to effectively present complex data structures in an intuitive manner. This not only enhances the user experience but also ensures that our applications are functional and aesthetically pleasing.

As we gear up for the next journey, we will delve into another set of PrimeNG components that further elevate our application's interactivity and functionality. Get ready to explore navigation and layout components in the next chapter.

8

Working with Navigation and Layout Components

Navigating through a web application should be a seamless experience. How we structure our content, guide our users, and respond to their interactions can significantly influence their overall experience. This chapter dives deep into PrimeNG's navigation and layout components, designed to help us craft intuitive and user-friendly interfaces for our Angular applications.

In this exploration, we'll uncover the potential of PrimeNG's navigation components, understanding how they can be used to guide users through our application. From menus to breadcrumbs, and from tabs to accordions, we'll learn how to structure content, create navigation paths, and design responsive layouts that adapt to different screen sizes.

The overall goal of this chapter is to empower you with the knowledge and skills to leverage PrimeNG's navigation and layout components effectively. By the end of this chapter, you will be able to create seamless navigation experiences, organize content in a structured manner, and ensure responsive layouts that adapt to various devices. You will also have gained insights into handling navigation events and integrating them into your application's functionality.

In this chapter, we will cover the following topics:

- Introducing navigation and layout components
- Working with menus
- Introducing PrimeNG panels

Technical requirements

This chapter contains various code samples of PrimeNG displaying components. You can find the related source code in the `chapter-08` folder of the following GitHub repository: `https://github.com/PacktPublishing/Next-Level-UI-Development-with-PrimeNG/tree/main/apps/chapter-08`

Introducing navigation and layout components

In the realm of web development, creating intuitive and user-friendly interfaces is paramount to providing a seamless experience for users. Navigation and layout components play a pivotal role in achieving this goal. These components serve as the building blocks for organizing content, guiding users through different sections of an application, and ensuring responsive design.

What are navigation and layout components?

Navigation components provide users with a means to navigate through an application's various features, sections, and content. They offer intuitive and accessible ways to access different functionalities, improving the overall usability of an application. Examples of navigation components include menus, breadcrumbs, tabs, and toolbars.

On the other hand, **layout components** are responsible for structuring and organizing the presentation of content within an application. They ensure that information is displayed in a clear and visually appealing manner. Layout components provide the foundation for creating responsive and adaptive designs that can adapt to different screen sizes and devices.

Crafting an intuitive navigation and layout experience

Creating an intuitive navigation system is like designing a city's road network. It should be logical, easy to follow, and cater to the needs of its users. Here are some pointers that can help you to achieve that:

- *User-centric design*: Always design with the user in mind. Understand their needs, their habits, and their expectations. A navigation system that resonates with the user's intuition will always be more effective.
- *Consistency is key*: Whether it's the placement of navigation buttons or the style of drop-down menus, maintaining consistency across your application helps users build familiarity.
- *Feedback*: Provide feedback when users interact with navigation elements. Whether it's a button changing color when hovered over or a subtle animation when a menu opens, these small interactions can enhance the user experience.
- *Adaptability*: Ensure your navigation and layout components are responsive, adapting to different screen sizes and devices, and offering a seamless experience whether viewed on a desktop, tablet, or mobile.

Best practices for creating navigation and layout components

Here are some additional best practices to keep in mind when creating navigation and layout components:

- *Simplicity*: Overcomplicating navigation can confuse users; aim for clarity and ease of use.

- *Modularity*: Divide complex navigation and layout structures into smaller, reusable components. This promotes code reusability, maintainability, and scalability.
- *Accessibility*: Ensure that your navigation components are accessible to everyone, including those with disabilities. Use semantic HTML, provide alt text for images, and ensure components are keyboard navigable.
- *Testing*: Test your navigation components. This can be done through usability testing, where real users interact with your application. Their feedback can offer invaluable insights to improve your components.

Remember that these best practices serve as guidelines, and it is vital to adapt them to the specific needs and requirements of your application. Now, let's move on to another essential aspect of building intuitive UIs in Angular applications: PrimeNG menus.

Working with menus

Menus are a fundamental element of a UI that provide a navigational structure and allow users to access various features and functionalities of an application. In PrimeNG, you can find a wide range of menu components that can be easily integrated into your projects. In this section, we will explore what menus are, discuss when to use PrimeNG menus, and provide an example of using PrimeNG menus in an e-commerce application.

What are PrimeNG menus?

Menus can vary in complexity and design, ranging from simple text-based menus to more elaborate hierarchical menus with submenus and icons. PrimeNG offers several menu components that cater to different use cases and design requirements. Some popular menu components provided by PrimeNG include the following:

- `Menu`: The `p-menu` component is a versatile menu that supports various modes, such as popup, slide, and overlay. It can be used as a standalone menu or as a dropdown within other components.
- `Menubar`: The `p-menubar` component represents a horizontal menu bar commonly used in top-level navigation scenarios. It allows you to create a clean and concise navigation interface, especially useful for applications with multiple sections or modules.
- `MegaMenu`: The `p-megaMenu` component is designed for more complex navigation scenarios, allowing you to create multi-column menus with images, icons, and submenus.
- `ContextMenu`: The `p-contextMenu` component enables the display of context-specific menus that appear when users right-click or long-press on an element. It is useful for providing context-dependent actions or options.
- `TieredMenu`: The `p-tieredMenu` component is a hierarchical menu that supports multiple levels of nested menus. It is suitable for organizing options in a structured manner.

- `Breadcrumb`: The `p-breadcrumb` component is used to display a breadcrumb navigation trail that represents the user's current location within the application's hierarchy. It is typically placed near the top of the page and provides links to higher-level sections or pages.

These are just a few examples of the menu components available in PrimeNG. Depending on your application's requirements, you can choose the most appropriate menu component to create a seamless and intuitive navigation experience.

Creating a basic menu

Suppose you want a horizontal menu at the top of your e-commerce application's home page. This menu includes categories such as electronics, clothing, home and kitchen, and sports and fitness, and each category represents a drop-down menu that expands when users hover over it or click on it.

To get started, we need to import the necessary modules from the PrimeNG library:

```
import {MenuModule} from 'primeng/menu'
```

Once the dependencies are installed and imported, let's see how a PrimeNG menu can be set up:

```
import { MenuItem } from 'primeng/api'

<p-menu [model]="menuItems" />

...

menuItems: MenuItem[] = [
    {
        label: 'Electronics',
        items: [
            { label: 'Computers', routerLink: '/products/computers' },
            { label: 'Smartphones', routerLink: '/products/smartphones'
},
            { label: 'Televisions', routerLink: '/products/televisions'
},
        ],
    },
    ...
]
```

Let's break down the code:

- `<p-menu [model]="menuItems" />`: This represents the usage of the PrimeNG `Menu` component. It binds the `model` property of the `Menu` component to the `menuItems` variable.

- `menuItems: MenuItem[]`: This is an array of `MenuItem` objects from the PrimeNG API. Each `MenuItem` object can have various properties, such as `label`, `items`, `routerLink`, `routerLinkActiveOptions`, and so on.

> **Note**
>
> The `routerLink` property in the menu items is an Angular feature that facilitates navigation between different routes in the application. The `routerLinkActiveOptions` property provides a means to verify and style the active menu by targeting the `p-menuitem-link-active` class. By default, the active class is applied to the route that matches the `routerLink` value defined in the `MenuItem` object. If you require different configurations, refer to the documentation available at `https://angular.io/api/router/IsActiveMatchOptions`.

In this example, each menu item is represented by a `label` property, which specifies the text to be displayed. The `items` property represents the submenu items for each category. Then the `routerLink` property is used to navigate to the corresponding product listing page when an item is clicked. Here is the result:

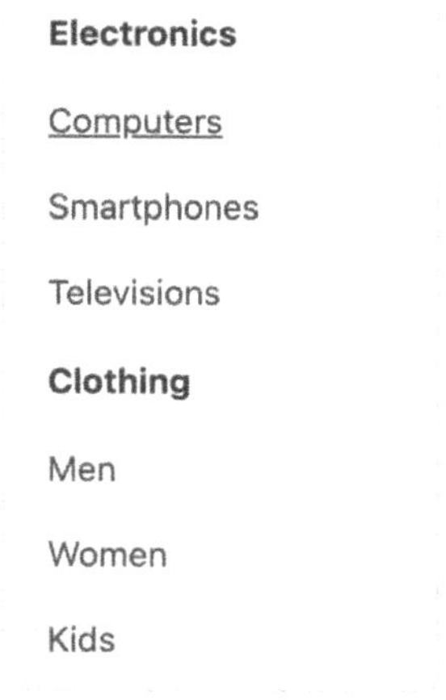

Electronics

Computers

Smartphones

Televisions

Clothing

Men

Women

Kids

Figure 8.1 – Basic menu

Working with Menubar

PrimeNG `Menubar` is a dynamic navigation component designed for horizontal layouts. It goes beyond a simple list of links, offering customization options to fit diverse application requirements. You can include links, buttons, and other UI components within `Menubar`, making it versatile and adaptable.

To get started, we need to import the necessary modules from the PrimeNG library:

```
import { MenubarModule } from 'primeng/menubar'
```

After that, we can create a menu bar by utilizing the PrimeNG `p-menubar` component:

```
<p-menubar [model]="menuItems" />
```

Let's take a look at the result:

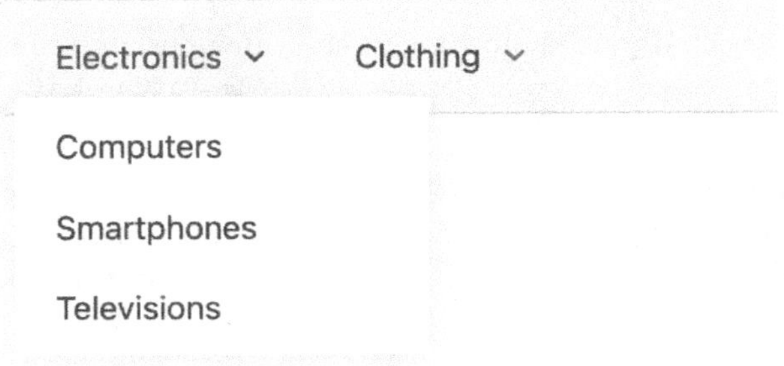

Figure 8.2 – Menubar

In this example, `Menubar` hosts two main sections: **Electronics** and **Clothing**. The **Electronics** section is further broken down into sub-categories such as **Computers**, **Smartphones**, and **Televisions**. This structure ensures that users can swiftly navigate to their desired product category or access their account settings without any hassle.

Working with MegaMenu

PrimeNG `MegaMenu` is a drop-down navigation component that displays submenus in a two-dimensional panel, which is ideal for situations with extensive navigation choices, eliminating the need for scrolling through long lists. It proves particularly useful for websites or applications with multiple features or categories, providing organized and easily accessible navigation options.

To get started, we need to import the necessary modules from the PrimeNG library:

```
import { MegaMenuModule } from 'primeng/megamenu'
```

After that, we can create a menu bar by utilizing the PrimeNG `p-megaMenu` component:

```
<p-megaMenu [model]="megaMenuItems" />

...

megaMenuItems: MegaMenuItem[] = [
```

```
    {
      label: 'Categories',
      items: [
        [
          {
            label: 'Electronics',
            items: [
              { label: 'Laptops', routerLink: '/electronics/
laptops' },
              { label: 'Cameras', routerLink: '/electronics/
cameras' },
            ],
          },
          ...
        ],
      ],
    },
    ...
]
```

In this example, we define an array of `MegaMenuItem` objects to represent the menu structure. Each `MenuItem` object has a `label` property to specify the text displayed for the menu item. Additionally, the `items` property is used for nesting submenus within a menu item. The resulting `MegaMenu` component will have a top level named **Categories**:

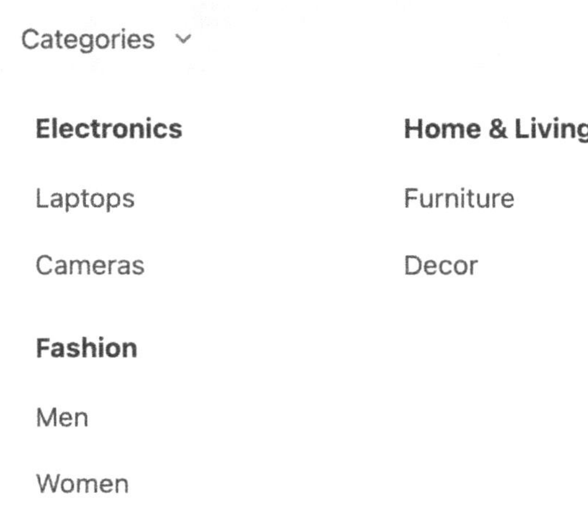

Figure 8.3 – MegaMenu

When users hover over **Categories**, a drop-down submenu will appear with additional subcategories and products. In this example, after hovering, you can see the **Electronics**, **Fashion**, and **Home & Living** subcategories. Within the **Electronics** subcategory, there are further options: **Laptops** and **Cameras**.

Working with ContextMenu

PrimeNG `ContextMenu` is a contextual menu component that pops up in the UI upon a specific user action, typically a right-click. It provides a list of actions that users can perform, which are contextually relevant to the area or the element they've interacted with. Instead of navigating to the top menu or searching for options, the context menu brings the actions directly to the user.

A context menu is used when you want to offer users quick actions without cluttering the UI. It's especially beneficial in the following cases:

- The screen space is limited
- You want to provide options that are relevant to a specific element or area
- You aim to reduce the number of clicks a user has to make

For instance, in a text editor, right-clicking might bring up options to cut, copy, or paste. In a photo viewer, it might offer options to zoom, save, or share.

Let's envision an e-commerce platform where users can browse products. A context menu can provide quick actions for these products. To get started, we need to import the necessary modules from the PrimeNG library:

```
import { MegaMenuModule } from 'primeng/megamenu'
```

After that, let's create a context menu for a product image:

```
<img
    #img
    src="assets/placeholder.png"
    alt="Product"
    aria-haspopup="true"
    class="max-w-full"
/>
<p-contextMenu [target]="img" [model]="contextMenuItems" />

...

contextMenuItems: MenuItem[] = [
    {
        label: 'View Details',
        icon: 'pi pi-search',
```

```
        command: (event) => this.viewProduct(event.item),
      },
      {
        label: 'Add to Cart',
        icon: 'pi pi-shopping-cart',
        command: (event) => this.addToCart(event.item),
      },
      {
        label: 'Add to Wishlist',
        icon: 'pi pi-heart',
        id: 'wishlist',
        command: (event) => this.addToWishlist(event.item),
      },
  ]

  viewProduct(item: MenuItem) {
    // Logic to view product details
  }

  addToCart(item: MenuItem) {
    // Logic to add product to cart
  }

  addToWishlist(item: MenuItem) {
    // Logic to add product to wishlist
  }
```

Let's break down the code:

- `<img #img .../>`: This is a template reference variable used to reference the `<img>` element within the Angular component
- `<p-contextMenu [target]="img" [model]="contextMenuItems" />`: This part of the code defines the PrimeNG `ContextMenu` component and configures its properties:
  - `[target]="img"`: This binds the target property of the `ContextMenu` component to the `img` template reference variable. This means that the `ContextMenu` component will be triggered when the user right-clicks on the referenced `<img>` element.
  - `[model]="contextMenuItems"`: This binds the `model` property of `ContextMenu` to the `contextMenuItems` array.

- `contextMenuItems: MenuItem[]`: This is an array of `MenuItem` objects. Each object represents a menu item in the context menu. It has properties such as `label` (the text displayed for the menu item), `icon` (an icon associated with the menu item), and `command` (a function to be executed when the menu item is selected). In this case, the `command` properties are set to call specific methods (`viewProduct()`, `addToCart()`, and `addToWishlist()`) in the Angular component.

Let's take a look at the result:

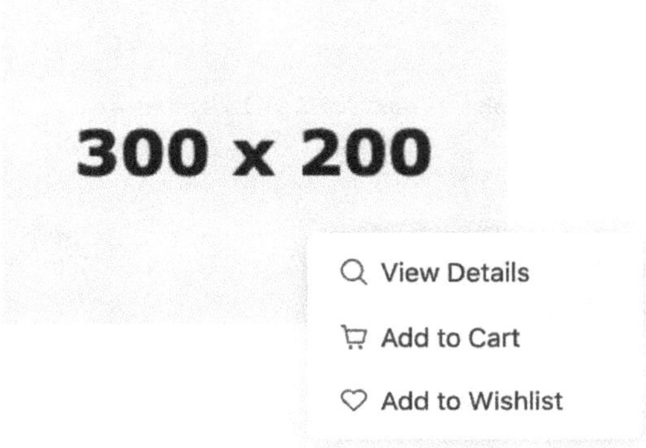

Figure 8.4 – Example of ContextMenu

In this setup, when a user right-clicks on a product, they're presented with options to view its details, add it to the cart, or add it to their wishlist. This enhances the shopping experience, making actions swift and straightforward.

Working with TieredMenu

PrimeNG `TieredMenu` is a multilevel menu system that allows options to be organized in a hierarchical structure. Instead of a flat list, you get a cascading menu where options can have child options, and those child options can have their own children, and so on. This tiered structure is visually intuitive, allowing users to navigate through categories and subcategories with ease.

PrimeNG `TieredMenu` is an excellent choice in various scenarios where hierarchical navigation menus are required. Here are some situations where you can leverage the power of `TieredMenu`:

- *Complex application menus*: When you have an application with a large number of menu items and submenus, `TieredMenu` simplifies the management and organization of the menu structure. It allows you to create a logical hierarchy of menus, making it easier for users to navigate through the application.

- *E-commerce websites*: `TieredMenu` is particularly useful in e-commerce websites that often have extensive product categories and subcategories. By using `TieredMenu`, you can create a user-friendly navigation system that enables shoppers to browse through different product categories and subcategories effortlessly.
- *Admin dashboards*: Admin dashboards typically have multiple sections and sub-sections, each requiring its own set of menus. `TieredMenu` provides a clean and organized way to represent these menus, allowing administrators to access various functionalities and settings with ease.
- *Multilevel dropdowns*: If you need to implement multilevel drop-down menus, `TieredMenu` simplifies the process. It handles the complexity of managing nested menus and ensures smooth transitions between different levels of hierarchy.

To illustrate the usage of PrimeNG `TieredMenu` in an e-commerce context, let's consider a scenario where we have an online store selling electronic devices. We want to create a navigation menu that allows users to browse through different product categories and subcategories.

To get started, we need to import the necessary modules from the PrimeNG library:

```
import { TieredMenuModule } from 'primeng/tieredmenu'
```

After that, we can enable PrimeNG `TieredMenu` by adding the following code:

```
<p-tieredMenu [model]="tieredMenus" />

...

tieredMenus: MenuItem[] = [
  {
    label: 'Electronics',
    icon: 'pi pi-tablet',
    items: [
      {
        label: 'Computers',
        icon: 'pi pi-desktop',
        items: [
          { label: 'MacBook Air', icon: 'pi pi-apple' },
          { label: 'Ultrabooks', icon: 'pi pi-desktop' },
          { label: 'Mobile Workstations', icon: 'pi pi-mobile' },
        ],
      },
      ...
    ],
  },
]
```

Let's break down the code:

- `<p-tieredMenu [model]="tieredMenus" />`: This represents the PrimeNG `TieredMenu` component. The `model` attribute binds the `tieredMenus` property in the component class to the `TieredMenu` component.
- `tieredMenus: MenuItem[]`: This is an array of `MenuItem` objects from the PrimeNG API. Each `MenuItem` object can have various properties, such as`label`, `icon`, `items`, `routerLink`, and so on.

Let's have a look at the result:

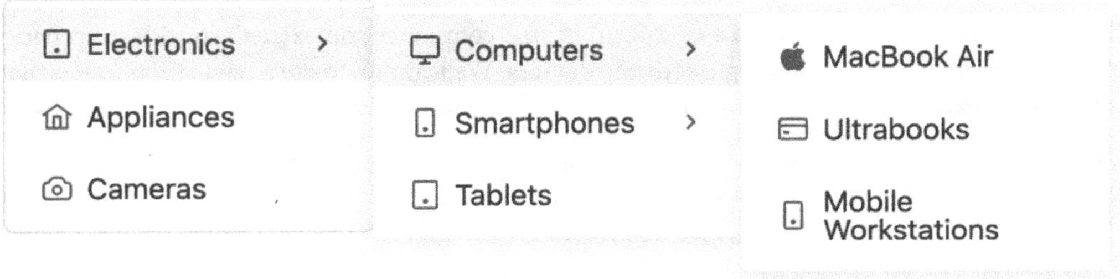

Figure 8.5 – A tiered menu

In this example, we have a top-level menu item, **Electronics**, with several submenus: **Computers**, **Smartphones**, and **Tablets**. Each submenu can have its own submenus, creating a hierarchical structure. Users can navigate through the menu items and submenus by hovering over the respective items.

Working with Breadcrumb

PrimeNG `Breadcrumb` is a navigation component that indicates the current location within an application's hierarchical structure. It provides a series of links, each representing a level in the hierarchy, leading back to the home page or main dashboard.

> **Note**
>
> Breadcrumbs should be used as a secondary navigation aid – they complement the main navigation but shouldn't replace it. For this reason, ensure that your application also has a primary navigation system, such as a menu or sidebar.

To get started, we need to import the necessary modules from the PrimeNG library:

```
import { BreadcrumbModule } from 'primeng/breadcrumb'
```

After that, we can enable PrimeNG Breadcrumb by adding the following code:

```
<p-breadcrumb [model]="breadcrumbItems" />

...

breadcrumbItems: MenuItem[] = [
    { icon: 'pi pi-home', routerLink: '/' },
    { label: 'Electronics', routerLink: '/electronics' },
    {
       label: 'Computers',
       routerLink: '/electronics/computers',
    },
    { label: 'MacBook Air', routerLink: '/electronics/computers/
macbook-air' },
]
```

Let's break down the code and explain each part:

- <p-breadcrumb [model]="breadcrumbItems" />: This represents the actual usage of the PrimeNG Breadcrumb component. The [model] attribute is used to bind the breadcrumbItems array to the component's model property.
- breadcrumbItems: MenuItem[]: This is an array of MenuItem objects from the PrimeNG API. Each MenuItem object can have various properties such as label, icon, items, routerLink, and so on.

Let's have a look at the result:

Figure 8.6 – Breadcrumb

In this example, the breadcrumb starts with the Home link, which takes users back to the website's home page. The subsequent links represent the category hierarchy: **Electronics**, **Computers**, and finally, the current page, **MacBook Air**.

Having delved into the diverse capabilities of PrimeNG's menus, we've seen how they can streamline navigation and enhance the user experience in applications. Now, let's transition to exploring PrimeNG panels, which are essential components that offer a versatile way to organize and present content within an application.

Introducing PrimeNG panels

Panels in PrimeNG serve as containers for content, allowing you to structure and group information in a visually appealing way. Each panel type has its unique characteristics and features, catering to different use cases within an application. Some popular PrimeNG panels include `Panel`, `ScrollPanel`, `Accordion`, `Splitter`, `Fieldset`, and `TabView`.

PrimeNG panels can be used in various scenarios where content organization and presentation are crucial. Here are some situations where you can benefit from using PrimeNG panels:

- **Collapsible and expandable sections**: Panels such as `Accordion` are useful when you have multiple sections of content and want to conserve space by allowing users to expand only the sections that they are interested in
- **Scrollable content**: When dealing with content that exceeds the available space, the `ScrollPanel` component enables users to scroll through the content, ensuring that all information remains accessible
- **Resizable and collapsible panels**: The `Splitter` component is valuable when you need to create resizable and collapsible panels, enabling users to customize the layout according to their preferences
- **Grouping related fields**: The `Fieldset` component is particularly useful when you have a form with related fields that need to be visually grouped together, improving the user experience and understanding
- **Tabbed navigation**: The `TabView` component is beneficial when you have multiple sets of related information or functionalities that can be organized into tabs, allowing users to switch between them easily

To get started, let's create a basic PrimeNG panel.

Creating a basic panel

A PrimeNG panel is essentially a container that wraps around content, providing it with a structured appearance. It comes with an optional header, which can be used to give a title or context to the content inside. The beauty of the panel lies in its simplicity. It doesn't impose any specific style or behavior on the content; instead, it offers a neat boundary, making the content stand out.

Whether you're designing a dashboard, a form, or a content page, the panel can be your go-to component to give structure to your content. It's especially useful when you want to group related pieces of information together, making it easier for users to process and understand.

> **Note**
>
> While PrimeNG `Panel` is versatile, it's essential to use it judiciously. Overusing it can make a page look cluttered. Always aim for a balance between design and functionality.

Let's dive into the realm of e-commerce to see PrimeNG `Panel` in action. In an e-commerce website, the `Panel` component can be used to display detailed information about a product. For example, when a user clicks on a product thumbnail or name, a panel can slide in or expand to reveal the product details, including images, descriptions, specifications, and customer reviews. This allows users to explore the product information without leaving the current page.

To get started, we need to import the necessary modules from the PrimeNG library:

```
import { PanelModule } from 'primeng/panel'
```

Once the dependencies are installed and imported, let's see how a PrimeNG panel can be set up:

```
<p-panel header="Product Details" [toggleable]="true">
    <p>
        Experience the power of Laptop XYZ, featuring the latest
processor and a
        sleek design.
    </p>
    <!-- other information -->
</p-panel>
```

Let's break down the code and explain each part:

- `<p-panel>`: This is the main component tag for creating a panel in PrimeNG.
- `header="Product Details"`: This attribute sets the title or header of the panel to `"Product Details"`.
- `-toggleable]="true"`: This attribute makes the panel's content collapsible. When set to `true`, it allows users to click on the panel header to toggle (show/hide) the content inside the panel.

Let's take a look at the result:

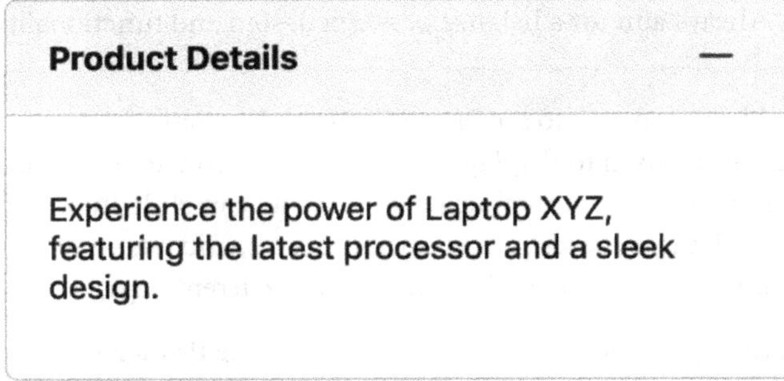

Figure 8.7 – Basic panel

In this example, you can expand or collapse the product details by clicking on the minus or plus icon.

Working with ScrollPanel

PrimeNG `ScrollPanel` enables you to create scrollable areas within their applications, allowing users to view content that exceeds the available space. The `ScrollPanel` component provides a native-like scrolling experience, which means that when users navigate through the app, the scrolling feels smooth and natural, just like it would on the device's built-in applications, and the component also supports both horizontal and vertical scrolling.

Unlike the default browser scrollbars, which can be inconsistent across different platforms, `ScrollPanel` offers a uniform look and feel. It's not just about aesthetics; it's about providing a smoother, more intuitive scrolling experience for users.

Let's improve the basic panel shown in *Figure 8.7* by adding `ScrollPanel`. To start, import the necessary modules from the PrimeNG library:

```
import { ScrollPanelModule } from 'primeng/scrollpanel'
```

Now, let's see how PrimeNG `ScrollPanel` can be set up:

```
<p-panel header="Product Details" [toggleable]="true">
   <p-scrollPanel [style]="{ width: '100%', height: '200px' }">
      <p>
         Lorem ipsum dolor sit amet, consectetur adipiscing elit.
Quisque ut...
      </p>
      <!-- Other content -->
   </p-scrollPanel>
</p-panel>
```

Let's break down the code and explain its functionality:

- `<p-panel ...>`: This represents the PrimeNG `Panel` component.
- `<p-scrollPanel [style]="{ width: '100%', height: '200px' }">`: This creates a scrollable viewport within the panel to handle overflowing content. It allows users to scroll through the content vertically. The `[style]` attribute is used to define the dimensions of the scrollable area. In this case, the width is set to `100%`, and the height is set to `200px`.

Let's take a look at the result:

Product Details
ipsum a iaculis. Interdum et malesuada fames ac ante ipsum primis in faucibus. Phasellus eget lobortis nisi. Donec nec accumsan tortor. Integer sit amet consequat justo, eu rhoncus diam. Proin dignissim non urna in dapibus. Proin ac tristique felis. Vivamus congue maximus eros. Aliquam egestas dictum leo, eu congue ipsum efficitur id.
Suspendisse posuere luctus nibh ac rutrum. Suspendisse feugiat purus eget quam cursus, a lobortis lorem tristique. In sollicitudin viverra urna, vel mattis nunc mollis vel. Nulla quis varius enim. Suspendisse auctor tempor quam, sed aliquam nunc tincidunt ut. Integer euismod felis ut blandit luctus. Proin vel sagittis lorem. Lorem ipsum dolor sit amet, consectetur adipiscing elit.

Figure 8.8 – ScrollPanel (with the scroll bar marked)

In this example, we defined `ScrollPanel` inside a basic panel with a fixed width and height. Since the content of the **Product Details** panel is quite large, the scrolling feature allows users to view the content effortlessly without losing the context of the entire page.

Working with Accordion

PrimeNG `Accordion` is a UI component that allows you to display content in a stacked manner. Think of it as a vertical stack of expandable/collapsible panels. Each panel has a title bar, and when you click on it, the content inside unfolds, revealing more details. This mechanism ensures that users are not overwhelmed with too much information at once. Instead, they can choose which sections to delve into, making their browsing experience more focused and less cluttered.

Accordions shine in scenarios where space is at a premium, and you need to present a list of items with associated detailed content. They're especially handy for the following circumstances:

- **Organizing related content**: Group related sections or topics, allowing users to quickly navigate to their area of interest

- **Forms with multiple steps**: Break down long forms into manageable chunks, guiding users step by step
- **FAQs**: Present a list of questions and expand to show answers
- **Product specifications**: In e-commerce, use accordions to display detailed specs or features of a product without overwhelming the user

Imagine you're developing an e-commerce platform that sells electronic gadgets. For each product, there's a wealth of information to convey specifications, user reviews, warranty details, and more. Using a PrimeNG accordion, you can neatly organize this information.

To get started, import the necessary modules from the PrimeNG library:

```
import { AccordionModule } from 'primeng/accordion'
```

Now we can see how a PrimeNG accordion can be set up:

```
<p-accordion>
    <p-accordionTab header="Specifications">
        <ul>
            <li>Processor: XYZ</li>
            <li>Memory: 8GB RAM</li>
            <!-- ... other specs ... -->
        </ul>
    </p-accordionTab>
    <p-accordionTab header="User Reviews">
        <p>"This product is fantastic! Highly recommend." - Alex G</p>
        <!-- ... other reviews ... -->
    </p-accordionTab>
    <p-accordionTab header="Warranty">
        <p>This product comes with a 2-year warranty covering...</p>
    </p-accordionTab>
    <!-- ... other tabs ... -->
</p-accordion>
```

The code provided demonstrates the usage of the PrimeNG `Accordion` component to create a set of tabs with different content sections. Let's break it down:

- `<p-accordion>...</p-accordion>`: This code snippet wraps the `Accordion` component around its child elements, indicating the start and end of the accordion section.
- `<p-accordionTab header="...">`: This element represents a section of the accordion. The header attribute defines the title of that section.

Let's have a look at the result:

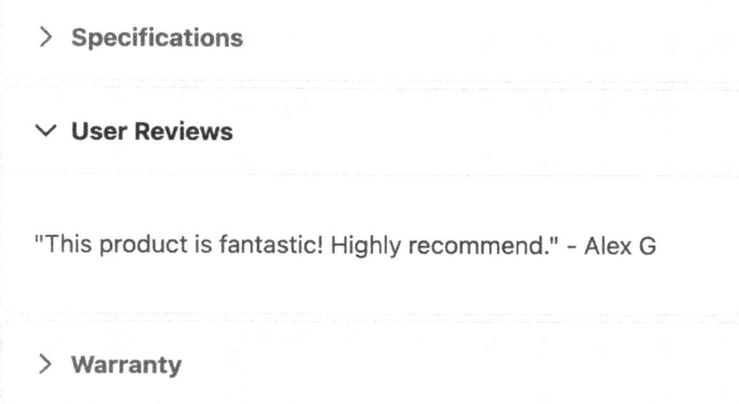

Figure 8.9 – Accordion

We have created three accordion tabs, so when users click on a title, the associated content inside the tab expands.

Working with Fieldset

PrimeNG `Fieldset` is a container component designed to group related content within a visually distinct boundary. It's similar to the classic HTML `<fieldset>` element but supercharged with additional features and styling. The most notable feature is its optional toggleable attribute, allowing users to expand or collapse the content within, making it perfect for sections that benefit from a hide/show functionality.

`Fieldset` is your go-to component for the following instances:

- **Grouping related elements**: This is especially the case for forms where you might want to group related input fields
- **Enhancing readability**: By segmenting content, you make it easier for users to process information
- **Interactive content presentation**: With its toggleable feature, you can present optional or supplementary information without overwhelming the main content

In the previous section, we displayed product information such as specifications, warranty details, and user reviews using PrimeNG `Accordion` (as seen in *Figure 8.9*). Instead of using `Accordion`, we can also use `Fieldset` to neatly package each section.

To get started, import the necessary modules from the PrimeNG library:

```
import { FieldsetModule } from 'primeng/fieldset'
```

Then, let's set up PrimeNG `Fieldset`:

```
<p-fieldset legend="Specifications" [toggleable]="true" class="block
mb-4">
    <ul>
        <li>Screen Size: 15.6 inches</li>
        <li>Processor: Intel i7</li>
        <li>RAM: 16GB</li>
        <!-- ... other specifications ... -->
    </ul>
</p-fieldset>

<p-fieldset legend="User Reviews" [toggleable]="true" class="block mb-
4">
    <p>"Fantastic product with great performance!" - Yen N</p>
    <!-- ... other reviews ... -->
</p-fieldset>

<p-fieldset
    legend="Warranty Details"
    [toggleable]="true"
    class="block mb-4"
>
    <p>
        This product comes with a 2-year warranty covering manufacturing
        defects.
        <!-- ... more warranty details ... -->
    </p>
</p-fieldset>
```

The provided code demonstrates the usage of the PrimeNG `Fieldset` component with specific attributes. Let's break down the code and explain each part:

- `<p-fieldset>`: The HTML tag that represents the PrimeNG `Fieldset` component.
- `legend="Specifications"`: The `legend` attribute sets the title or description of the `Fieldset` component. In this case, `legend` is set to `Specifications`, indicating that the content within the `Fieldset` component pertains to the specifications of a product.

- `[toggleable]="true"`: The `[toggleable]` attribute is a property binding that enables the toggleable functionality of the `Fieldset` component. When set to `true`, the content within the fieldset can be expanded or collapsed by the user. This allows the user to hide or reveal the specifications section as needed.

Let's take a look at the result:

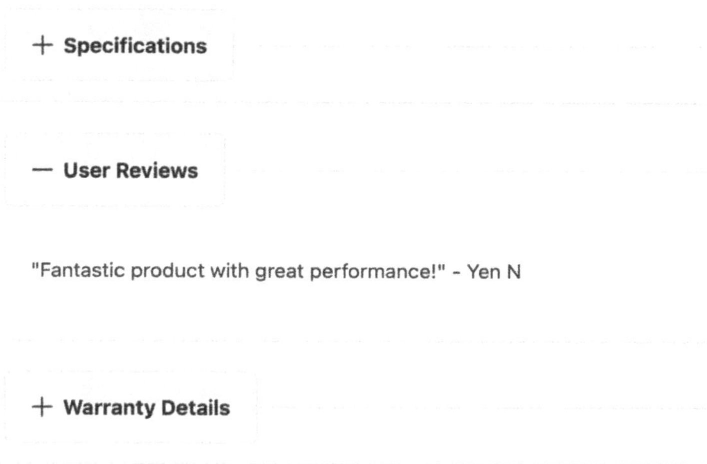

Figure 8.10 – Fieldset

In this example, each fieldset acts as a container for a specific type of information. Users can quickly glance at the legends (such as **Specification** or **User Review**) and decide which sections they want to delve into, expanding them as needed.

Working with TabView

PrimeNG `TabView` is a navigation component that allows you to display content broken down into multiple tabs. Each tab acts as a container for its unique content, ensuring that information is organized and easily accessible. With `TabView`, users can swiftly switch between different sections without being overwhelmed.

`TabView` is versatile and finds its place in various scenarios:

- **Settings or configuration pages**: This is so that different categories of settings can be grouped under individual tabs
- **Profile pages**: This is when you might want to separate user information, activity history, and settings into distinct tabs

- **Product descriptions in e-commerce**: This is when you want to segregate product details, customer reviews, and specifications
- **Documentation**: This is for when you want to separate guides, API references, and examples

In the previous sections, we displayed the product information using PrimeNG `Accordion` (*Figure 8.9*) and `Fieldset` (*Figure 8.10*). Now, let's try using `TabView`. To get started, import the necessary modules from the PrimeNG library:

```
import { TabViewModule } from 'primeng/tabview'
```

And now let's see how PrimeNG `TabView` can be set up:

```
<p-tabView>
    <p-tabPanel header="Warranty">
        <p>
            This product comes with a 2-year warranty covering
manufacturing
            defects.
            <!-- ... more warranty details ... -->
        </p>
    </p-tabPanel>
    <p-tabPanel header="Specifications">
        <ul>
            <li>Screen Size: 15.6 inches</li>
            <li>Processor: Intel i7</li>
            <li>RAM: 16GB</li>
            <!-- ... other specifications ... -->
        </ul>
    </p-tabPanel>
    <p-tabPanel header="Reviews">
        <p>"Spectacular product! Highly recommended" - Aaron D</p>
        <!-- ... other reviews ... -->
    </p-tabPanel>
</p-tabView>
```

Let's break down the code and explain each part:

- `<p-tabView>`: This PrimeNG component serves as the container for our tabs.
- `<p-tabPanel header="Warranty">`: These components represent different sections of product information inside the `tabView` container. The `header` attribute holds the value of the tab's label.

Let's have a look at the result:

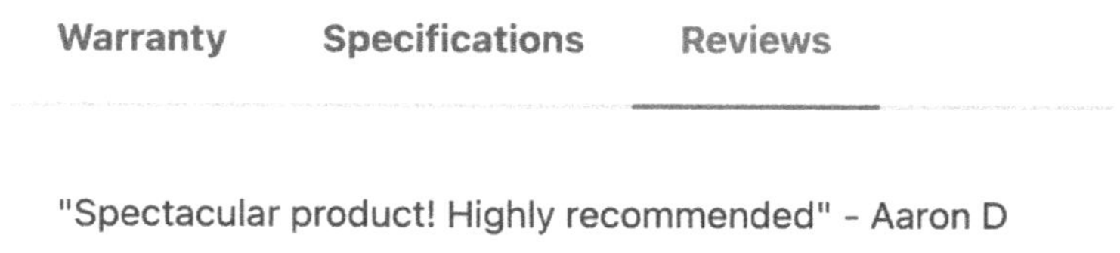

Figure 8.11 – TabView

By using PrimeNG `TabView`, we can create separate sections of the content that are displayed as tabs. Users can click on the tab headers (**Warranty**, **Specifications**, or **Reviews**) to switch between the corresponding sections and access the relevant information.

Working with Splitter

PrimeNG `Splitter` is a layout component that allows users to adjust the size of its child elements by dragging a divider. Think of it as a resizable partition that users can adjust according to their viewing preferences. It's particularly useful in scenarios where you want to provide an adjustable ratio between two or more content sections, be it horizontally or vertically.

The `Splitter` component shines in various scenarios:

- **Dashboard layouts**: You can use `Splitter` when you have multiple widgets or panels, and you want to give users the flexibility to adjust their sizes
- **Document editors**: The `Splitter` component can be used when you might have a code section on one side and a preview on the other
- **Image comparisons**: The `Splitter` component lets you allow users to adjust the view to compare two images side by side
- **Multi-panel interfaces**: The `Splitter` component can be used for applications where you want to offer adjustable multi-panel views

Let's consider an example where PrimeNG `Splitter` compares two images side by side. To start, import the necessary modules from the PrimeNG library:

```
import { SplitterModule } from 'primeng/splitter'
```

Then let's see how PrimeNG `Splitter` can be set up:

```
<p-splitter [style]="{ height: '300px' }" layout="horizontal">
  <ng-template pTemplate>
    <div class="col flex align-items-center justify-content-center">
```

```
            <img src="assets/placeholder.png" alt="Image 1" />
        </div>
    </ng-template>
    <ng-template pTemplate>
        <div class="col flex align-items-center justify-content-center">
            <img src="assets/placeholder.png" alt="Image 2" />
        </div>
    </ng-template>
</p-splitter>
```

Let's break down the code and explain each part:

- `<p-splitter>`: This represents the PrimeNG `Splitter` component.
- `[style]="{ height: '300px' }"`: The is used to apply inline CSS styles to the `Splitter` component. In this case, the height of the `Splitter` component is set to 300 px using the `height` property.
- `layout="horizontal"`: The `layout` attribute defines the orientation of the panels within the `Splitter` component. In this example, it is set to `horizontal`, indicating that the panels will be displayed side by side horizontally.
- `<ng-template pTemplate>`: The `pTemplate` directive is specific to PrimeNG and is used to mark the Angular template as a PrimeNG template.

Let's take a look at the result:

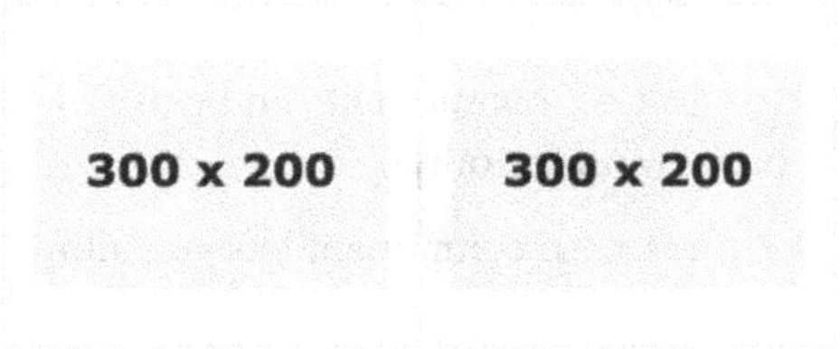

300 x 200 300 x 200

Figure 8.12 – Splitter

Here we created two panels within the `Splitter` component, with each panel containing an image for comparison purposes.

Having explored the various features and capabilities of PrimeNG panels, we have gained valuable insights into how this powerful component can enhance the layout and presentation of content in your Angular applications.

Summary

In this chapter, we navigated through the vast world of PrimeNG components. It took us on a journey through the intricacies of navigation and layout components. These elements are the backbone of any application, dictating how users interact with the content and ensuring a seamless experience.

We delved deep into various components from menus and panels. Each component serves a unique purpose, from organizing content to enhancing navigation. By now, you should have a firm grasp on when and how to use these components in your applications. These components will allow us to structure and present content in a meaningful way, improving the overall user experience. Moreover, through various examples, we've seen how these components can be integrated into real-world applications. These practical insights aim to bridge the gap between theoretical knowledge and actual implementation.

As we transition to the next chapter, we'll be diving into the art of customizing PrimeNG components with theming. Theming is a powerful tool that allows you to tailor the look and feel of components to align with brand guidelines or specific design preferences. We'll explore how to harness the power of theming to make PrimeNG components truly your own. So, gear up for a colorful journey into the world of customization and design!

Part 3: Advanced Techniques and Best Practices

In this part, you will delve into advanced techniques and best practices for working with PrimeNG. You will explore customization, optimization, reusability, internationalization, and testing strategies to enhance your PrimeNG-powered Angular applications.

By the end of this part, you will have a deep understanding of these advanced topics and be equipped with valuable skills to build robust, efficient, and user-friendly applications.

This part contains the following chapters:

9

Customizing PrimeNG Components with Theming

Every application has its own unique identity, and its visual appeal plays a significant role in defining that identity. While functionality is crucial, the look and feel of an application can significantly influence user experience. PrimeNG offers a robust theming system that allows you to tailor the appearance of components, ensuring that the application not only works well but also looks the part.

In this chapter, you will explore the process of customizing the appearance of PrimeNG components in Angular applications using theming. By mastering those techniques, you will gain the ability to tailor the visual presentation of PrimeNG components to align with the application's unique brand and design requirements.

We will also delve into various topics such as working with pre-built themes, creating custom themes, leveraging the Theme Designer tool, and overriding component styles. You will discover the power and flexibility of theming in PrimeNG and learn how to achieve a cohesive and personalized user interface.

The chapter will cover the following topics:

- Introducing PrimeNG theming
- Working with pre-built themes
- Creating your own custom themes
- Overriding component styles and other tips

Technical requirements

This chapter contains various working code samples on PrimeNG theming. You can find the related source code in the `chapter-09` folder of the following GitHub repository: `https://github.com/PacktPublishing/Next-Level-UI-Development-with-PrimeNG/tree/main/apps/chapter-09`.

Introducing PrimeNG theming

Theming plays a crucial role in creating visually appealing and consistent user interfaces, allowing you to customize the appearance of components to match the application's branding and design requirements. PrimeNG offers a comprehensive theming system that empowers you to create personalized and cohesive user interfaces.

PrimeNG theming isn't just about changing colors or fonts, though. It involves modifying various aspects of the component's appearance, such as colors, typography, spacing, and other design elements. PrimeNG provides a wide range of tools, resources, and guidelines to facilitate the theming process, enabling developers to create unique and visually appealing user interfaces.

PrimeNG theming is used in various scenarios, depending on the specific requirements of the application. Here are a few common use cases where PrimeNG theming proves valuable:

- **Branding and customization**: When you need to align the appearance of PrimeNG components with your application's branding guidelines, theming allows you to create a consistent and personalized look and feel.
- **Application-specific design**: In some cases, the default styles of PrimeNG components may not align with the specific design requirements of your application. Theming enables you to modify the appearance of components to match your application's visual design language, ensuring a cohesive and harmonious user interface.
- **Consistent styling**: When building large-scale applications with multiple developers or teams, theming ensures consistency in the visual styles across different components. By adhering to a unified theming approach, you can maintain a coherent user experience throughout the application.

Having grasped the foundational knowledge of PrimeNG theming, it's time to delve into the practical aspects. One of the quickest ways to kickstart your theming journey is by leveraging PrimeNG's vast array of pre-built themes, which we'll explore next.

Working with pre-built themes

PrimeNG **Pre-Built Themes** are a collection of predefined style sheets that define the visual appearance of PrimeNG components. These themes are based on popular design frameworks such as Bootstrap and Material Design, and they come with a wide range of color schemes and variations. Each theme provides consistent styling for all PrimeNG components, ensuring a cohesive and polished look for your application.

The pre-built themes are shipped with PrimeNG as part of the `npm` distribution, are easily importable, and can be applied to your application with a few simple steps. These themes are also highly customizable, allowing you to tweak the colors, fonts, and other visual properties according to your project's requirements.

When to use PrimeNG pre-built themes

While customization offers a unique identity, there are scenarios where pre-built themes can really save the day:

- **Rapid prototyping**: When you're in the initial stages of application development and need a quick design to visualize the functionality
- **Consistent design language**: For projects where a consistent design across multiple applications or modules is essential
- **Reduced development time**: When the project timeline is tight, and there's no room for extensive design iterations

Example of PrimeNG pre-built themes

In previous chapters, we had a chance to work with PrimeNG theming. In this chapter, let's refresh our minds on how to integrate PrimeNG theming into our Angular application.

To use a pre-built theme, it's as simple as importing it into your project. Let's see how:

1. Navigate to the PrimeNG Built-in Themes (`https://primeng.org/theming#builtinthemes`) and pick a theme that resonates with your project's vibe.
2. Once you've chosen a theme, integrate it into your project. For instance, if you've selected the `lara-light-blue` theme, add the following lines to `styles.scss` or `styles.css`:

   ```
   //styles.scss

   @import 'primeng/resources/themes/lara-light-blue/theme.css';
   @import 'primeng/resources/primeng.css';
   ```

   This import statement ensures that the styles defined in the `lara-light-blue` theme are applied to your application. Once the theme is imported, all PrimeNG components will automatically adopt the styling defined by the theme.

3. Next, you can start using PrimeNG components in your application, such as the `p-button`, `p-card`, and `p-table` components. These components will inherit the styles defined by the pre-built theme, giving them a consistent and visually appealing appearance.

> **Note**
>
> Your applications might have custom fonts or stylings. Always ensure that the theme's CSS is loaded after the default PrimeNG CSS to ensure the theme styles take precedence.

We have explored the convenience of PrimeNG's pre-built themes, which offer a wide selection of visually appealing styles out of the box. Now, let's dive into the topic of switching themes, where we'll learn how to seamlessly switch between different PrimeNG themes to suit our application's design requirements and preferences.

Switching themes

One of the key advantages of PrimeNG pre-built themes is that you can switch between different themes on the fly. This feature allows users of your application to choose their preferred theme, providing them with a personalized and customizable experience.

To switch themes, you typically need to replace the theme CSS file reference in your project. For instance, if you are currently using the `lara-blue-light` theme and wish to switch to the `bootstrap4-light-purple` theme, you'd replace the CSS file reference in your `index.html` file.

Here is a step-by-step guide on how to do it:

1. In order to switch themes, you need to have all your themes ready in the `assets` folder. You can find a full list of the built-in PrimeNG themes under the `node_modules/primeng/resources/themes` folder:

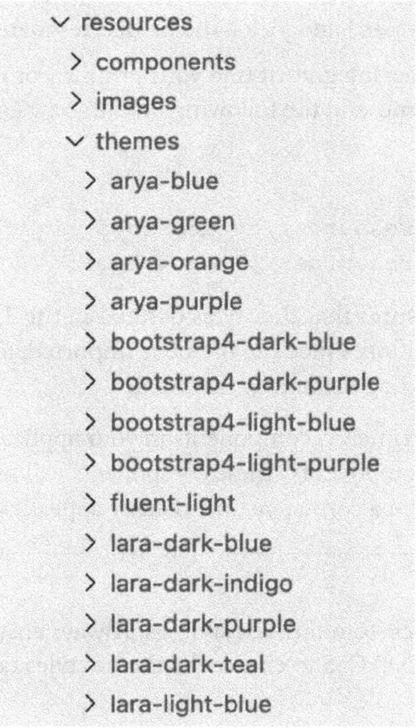

Figure 9.1 – PrimeNG built-in themes

After that, you can copy the themes that you want to use in your application to your `assets` folder:

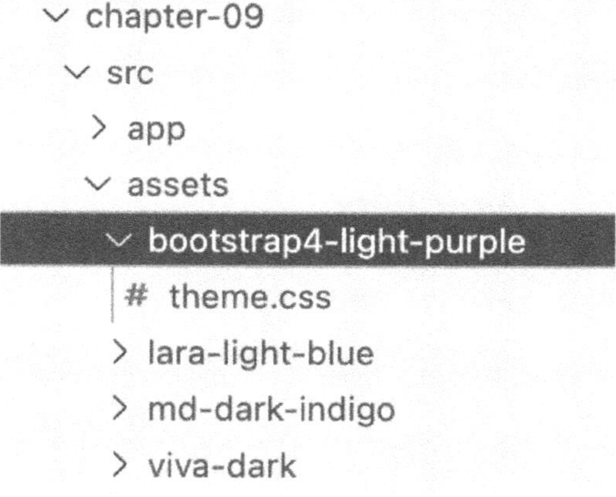

Figure 9.2 – Copied built-in themes

In this example, I copied four built-in themes to our application: `bootstrap4-light-blue`, `lara-light-blue`, `md-dark-indigo`, and `viva-dark`.

2. Add your default theme reference to `index.html`. So, instead of having your default in the `style.scss` file, you will have it in the `index.html` file:

```
<!DOCTYPE html>
<html lang="en">
  <head>
    <meta charset="utf-8" />
    <title>chapter-09</title>
    <base href="/" />
    <meta name="viewport" content="width=device-width, initial-
scale=1" />
    <link rel="icon" type="image/x-icon" href="favicon.ico" />
    <link
      id="theme-link"
      rel="stylesheet"
      type="text/css"
      href="assets/lara-light-blue/theme.css"
    />
  </head>
  <body>
    <primengbook-root />
  </body>
</html>
```

In this code, we added a reference to the `lara-light-blue` theme with the `theme-link` ID.

3. Now, it's time to add switch-theme functionality to our component. Here is the code:

```
<div class="flex align-items-center" *ngFor="let theme of
themes">
  <p-radioButton
    [name]="theme.name"
    [value]="theme.value"
    [(ngModel)]="selectedTheme"
    inputId="{{ theme.value }}"
    (onClick)="changeTheme()"
  />
  <label for="{{ theme.value }}" class="ml-2">{{ theme.name }}</
label>
</div>

...

themes = [
  { name: 'Lara Light Blue', value: 'lara-light-blue' },
  { name: 'Bootstrap4 Light Purple', value: 'bootstrap4-light-
purple' },
  { name: 'Viva Dark', value: 'viva-dark' },
  { name: 'Material Dark Indigo', value: 'md-dark-indigo' },
]

selectedTheme = 'lara-light-blue'

changeTheme() {
  const themeLink = document.getElementById('theme-link')
  themeLink?.setAttribute('href', `assets/${this.selectedTheme}/
theme.css`)
}
```

Let's break down the code:

- `<div class="flex align-items-center" *ngFor="let theme of themes">`: This code iterates over the `themes` array using the `ngFor` directive. For each theme in the array, it creates a radio button input using the `p-radioButton` component.
- `themes = [...]`: This defines the `themes` array, which contains objects representing different themes. Each theme object has a `name` property that represents the display name of the theme and a `value` property that represents the unique identifier of the theme.
- `selectedTheme`: This indicates the initially selected theme value.
- `changeTheme() {...}`: This method is called when a radio button is clicked. It retrieves the `link` element with the `theme-link` ID. After that, it will update the `href` attribute of the `link` element to point to the CSS file of the selected theme.

Let's take a look at the result:

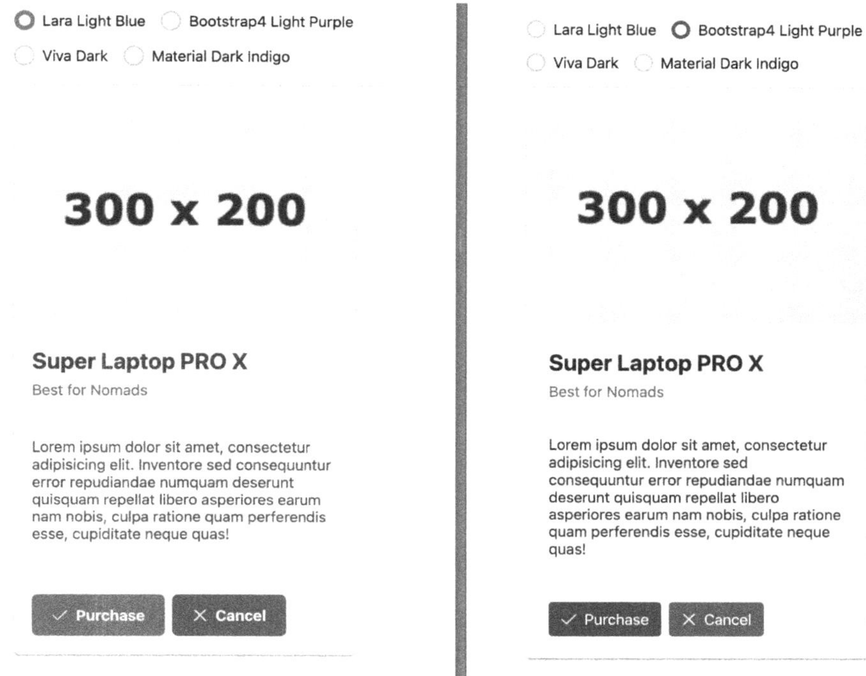

Figure 9.3 – Switching themes

We have set up a theme switcher that allows users to select a theme using radio buttons. When a radio button is clicked, we will be able to switch to the desired theme. You can see that on the left we're using the `Lara Light Blue` theme, and on the right, we're using `Bootstrap4 Light Purple`, which will give a different experience to our users.

In general, working with pre-built themes in PrimeNG provides you with a convenient and efficient way to style applications. In the next section, we will explore the process of creating custom themes in PrimeNG, empowering you to craft a truly personalized and distinctive user experience.

Creating your own custom themes

While PrimeNG offers plenty of pre-built themes, there might be situations where you want a unique look and feel that aligns more closely with your brand or specific design requirements. That's where custom themes come into play. These themes allow you to tailor the appearance of PrimeNG components to your exact specifications.

What are PrimeNG custom themes?

A **custom theme** in PrimeNG is essentially a set of CSS styles that override the default styles of PrimeNG components. By creating a custom theme, you have the flexibility to define colors, fonts, spacings, and other design elements that match your brand's identity or the specific design language of your project.

When are PrimeNG custom themes used?

Custom themes are particularly beneficial in the following situations:

- You're building a branded application where the visual identity needs to be consistent with other digital assets of your brand
- The pre-built themes don't align with the specific design requirements of your project
- You're aiming for a unique user interface that stands out from typical applications
- There's a need to adhere to specific accessibility guidelines that might not be covered by the default themes

How to create PrimeNG custom themes

Creating a custom theme might sound daunting, but with PrimeNG's structure, it's quite straightforward. You have three alternatives to choose from, each offering its own advantages and flexibility. Let's delve into each option:

- **Visual Editor**: The Visual Editor is a user-friendly tool provided by PrimeNG that allows you to visually customize and style your theme
- **Command-line Sass compilation**: If you prefer a more hands-on approach, you can choose to compile your theme using the command-line Sass tool
- **Embedding SCSS files in your project**: The third alternative involves embedding the SCSS files directly within your project's directory structure

> **Note**
>
> As of the book's release, the Visual Editor is currently disabled. However, there is good news - the PrimeTek team has made the decision to open-source the Designer, making it freely available to use. The anticipated release for this exciting development is scheduled for Q1 2024. For the most up-to-date information, please visit the official PrimeTek website.

In all three options, it is crucial to import the generated theme file into your project. This ensures that the customized theme is properly applied to the PrimeNG components, allowing you to enjoy the benefits of your personalized visual style.

Choose the option that aligns best with your preferences and project requirements, and embark on the journey of creating a unique and visually appealing theme for your PrimeNG-powered application.

Creating a custom theme via the Visual Editor

The world of theming in PrimeNG has been revolutionized with the introduction of the **Visual Editor**. No longer do you need to dive deep into lines of Sass or CSS to get the perfect look for your components. With the Visual Editor, creating a custom theme is as intuitive as dragging a slider or picking a color from a palette.

With its intuitive interface, you can modify various aspects of the theme, such as colors, typography, spacing, and more. The Visual Editor provides real-time previews, making it easy to see the changes as you customize your theme. Once you are satisfied with the modifications, you can export the theme file, which can be directly imported into your project.

Creating a custom theme with the Visual Editor is a breeze. Here's a step-by-step guide:

1. Navigate to the PrimeNG theming page and launch the Visual Editor.
2. Start with a pre-built theme that's closest to your desired look. This serves as a foundation upon which you can make further customizations. In this example, I will choose the `lara-light` theme.

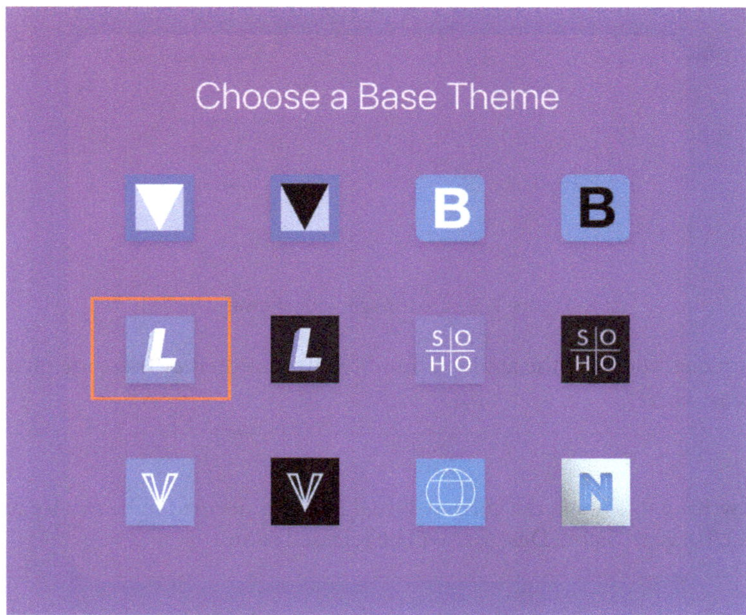

Figure 9.4 – Choose a base theme

3. Use the Visual Editor's intuitive controls to adjust colors, fonts, and other design elements. As you make changes, you'll see a real-time preview of how components look.

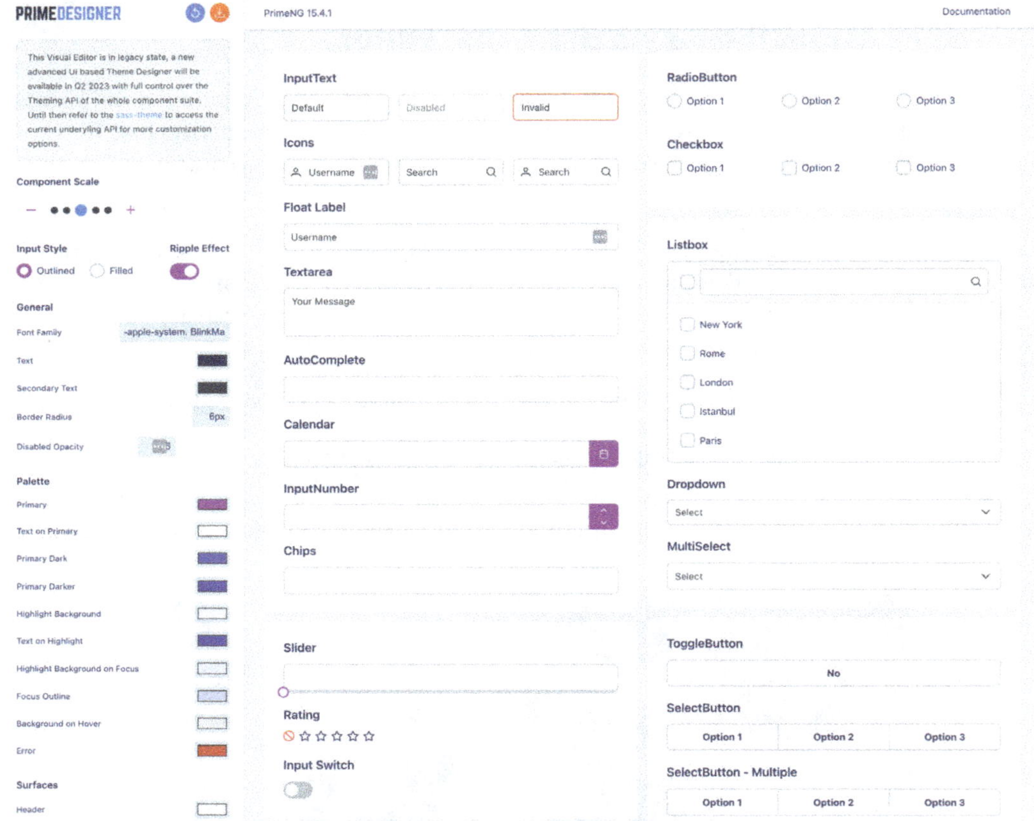

Figure 9.5 – Customize your theme

On the left panel, you have the ability to modify the base theme. There are plenty of options for you to customize.

> **Note**
>
> At the time of writing this book, the current Visual Editor is deemed to be in a legacy state. A new advanced UI-based Theme Designer will be released soon.

4. Once satisfied with your custom theme, simply export it. The Visual Editor will generate all the necessary CSS files for you.

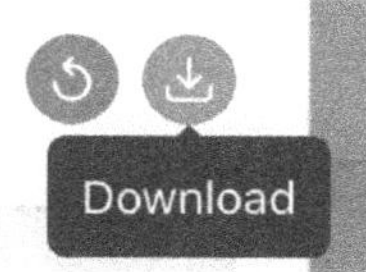

Figure 9.6 – Download your custom theme

After finishing your customization, you can start to download your custom theme by clicking on the **Download** button. Your custom theme will likely be under the `Download` folder with the name `theme.css`.

5. After downloading the exported files, you can start to include them in your Angular project. Make sure they're loaded after the default PrimeNG styles to ensure your customizations take precedence.

   If you only have one theme in your app, you can just put it in the `styles.scss` file or in `index.html`:

   ```
   // styles.scss

   @import 'assets/my-awesome-theme/theme.css';
   ```

Here is the result:

Figure 9.7 – Apply your custom theme

Now, your application has the desired look and feel of your custom theme.

Creating a custom theme via Sass compilation

The second method, using Sass compilation, gives you full control over the theme customization process. You can manually edit the theme's SCSS files, adjusting variables to achieve the desired visual style. Once you have made the necessary modifications, you can use the command-line Sass compiler to generate the CSS output. The compiled CSS file can then be imported into your project, ensuring that your custom theme is applied to the PrimeNG components.

Here is how we do it:

1. Clone the `primeng-sass-theme` repository from GitHub:

   ```
   git clone https://github.com/primefaces/primeng-sass-theme.git
   ```

2. Install the NPM packages:

   ```
   cd primeng-sass-theme
   npm install
   ```

3. After installing the necessary packages, we can find all of the SCSS files under the `themes` | `mytheme` directory:

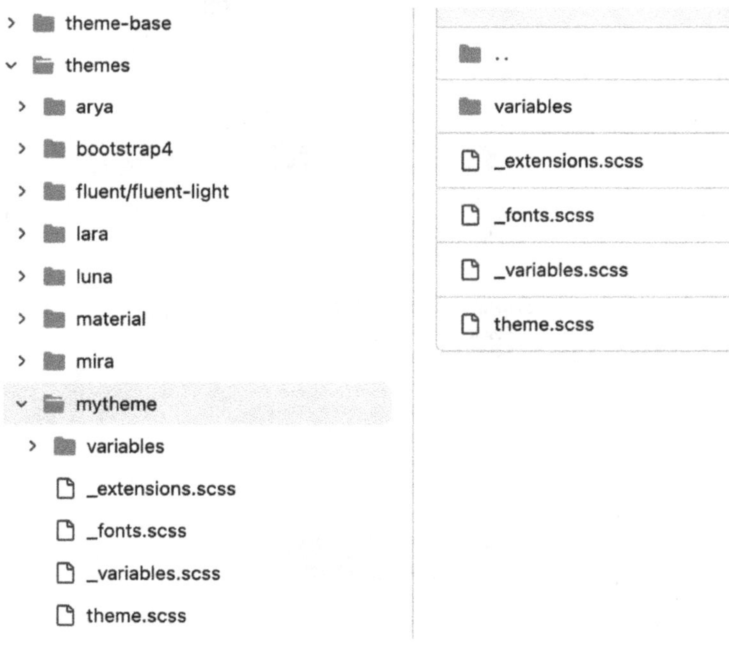

Figure 9.8 – mytheme folder

Under the mytheme folder, you have a few options to customize your custom theme:

- variables: You can customize the CSS variables under this folder, changing the button background, primary text color, border radius, and so on
- _extension.scss: Update this file if you want to override the component designs
- _font.scss: This is where you can define a custom font for your project
- theme.scss: This file imports theme files, along with the theme-base folder, in order to combine everything

4. After making changes to your custom theme, compile your changes by running the following command:

```
sass --update themes/mytheme/theme.scss:themes/mytheme/theme.css

[2023-09-17 14:33:46] Compiled themes/mytheme/theme.scss to
themes/mytheme/theme.css.
```

You can see that, after compiling, we created a theme.css file under the mytheme folder, which now can be added to our project.

Creating a custom theme by embedding SCSS files

This third approach allows you to integrate the theme customization seamlessly into your existing build environment. By placing the theme SCSS files in a designated location within your project, such as the assets or styles folder, you can leverage your build tools to automatically compile the SCSS files into CSS during the build process.

Here are the steps:

1. Copy the mytheme and theme-base folder to our assets folder. This option grants you the flexibility to incorporate the Angular CLI default process.

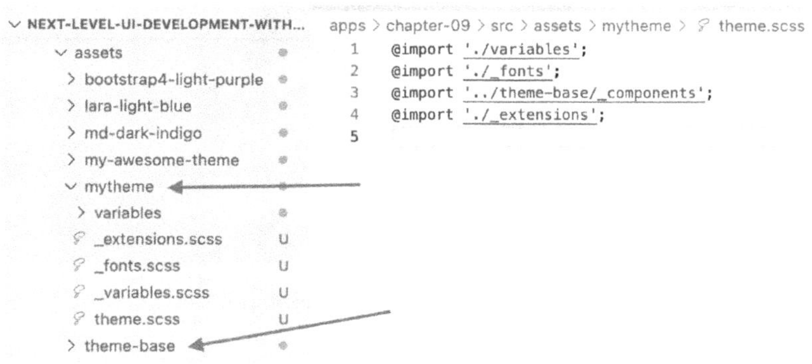

Figure 9.9 – mytheme in the Angular project

2. Then, all we need to do is import `theme.scss` from `mytheme` to `styles.scss`:

```
@import 'assets/mytheme/theme.scss';
```

3. Finally, when we modify anything under `mytheme`, the Angular CLI will handle the compilation for us naturally without any manual work.

In the journey of mastering PrimeNG theming, we've seen the power and flexibility of creating custom theme methods, making theme customization a straightforward and enjoyable process. Let's now shift our focus to overriding component styles and other advanced styling techniques to truly make our UIs stand out.

Overriding component styles and other tips

While PrimeNG offers plenty of themes and customization options, there will always be scenarios where we need to tweak certain styles to fit our application's unique requirements. This section will guide you through the process of overriding component styles and share some additional tips and tricks to enhance your theming experience.

How to override component styles

Overriding component styles in PrimeNG is similar to styling any other Angular component. The key is to understand the structure of the component you're trying to style and use specific CSS selectors. Here is how we do it:

1. **Inspect the component**: Before you can override a style, you need to know what you're targeting. Use your browser's developer tools to inspect the component and understand its structure. Here is an example of inspecting the browser:

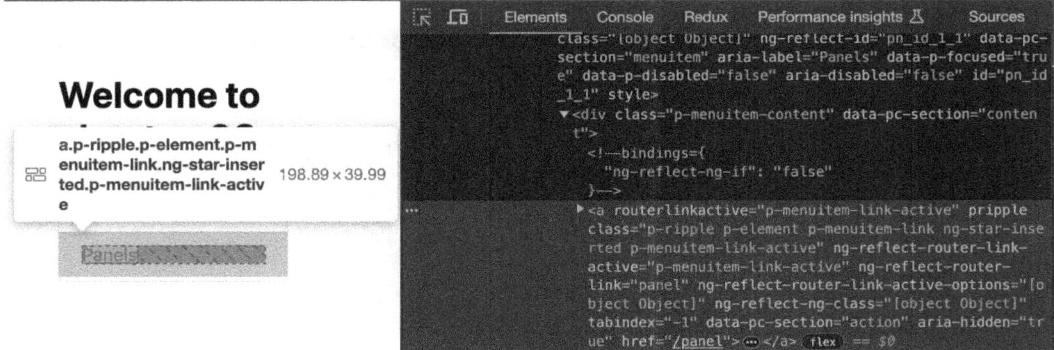

Figure 9.10 – Browser inspection example

In the screenshot provided, you can observe the detailed HTML element structure of the **Panels** menu located on the left side.

2. **Use specific selectors**: Once you've identified the element or class you want to override, write a more specific selector than the one used in the default PrimeNG styles. This ensures that your styles take precedence. For instance, in *Figure 9.10*, you can observe that the active menu is distinguished by the presence of the `p-menuitem-link-active` class, allowing you to selectively apply CSS styling to it.
3. **Use global styles**: You can add custom styling directly to `styles.scss` in order to apply styles to your component. For example, in order to apply underline to the active router link, you can add the following CSS to `style.scss`:

```
.p-menu .p-menuitem-link.p-menuitem-link-active {
  text-decoration: underline;
}
```

4. **Leverage ::ng-deep**: In Angular, the `::ng-deep` pseudo-class ensures that styles penetrate into child components. This is especially useful when trying to style the inner parts of a PrimeNG component. Here's an example:

```
:host ::ng-deep my-component .p-button {
    background-color: #333;
    color: #fff;
}
```

In this example, we can override the style of the PrimeNG button in our component, which makes it easier to locate and maintain. The changes only apply a different color and background color to our button, differentiating it from other buttons in your app.

Note

The Angular team decided to deprecate `::ng-deep` in future versions of Angular, so please use it sparingly. You can find more information about this at `https://angular.io/guide/component-styles#deprecated-deep--and-ng-deep`.

Note

Avoid using `!important`. While it might be tempting to use `!important` to force a style, it's a practice we should avoid. It makes future changes harder and can lead to unpredictable results.

Using PrimeNG common utils

PrimeNG **Utils** are a set of utility classes provided by PrimeNG to help you with common styling tasks. These utility classes offer a quick way to apply specific styles or behaviors to elements without having to write custom CSS. Here's a brief overview of some of the PrimeNG utils:

Class Name	Description
p-component	Applies component theming such as font family and font size to an element.
p-fluid	Applies 100% width to all descendant components.
p-disabled	Applies an opacity to display as disabled.
p-sr-only	Element becomes visually hidden, however accessibility is still available.
p-reset	Resets the browser's defaults.
p-link	Renders a button as a link.
p-error	Indicates an error text.

Figure 9.11 – Common PrimeNG util classes

Working with PrimeNG CSS variables

In modern web development, CSS variables (also known as CSS custom properties) have become a powerful tool for creating more flexible and maintainable style sheets. PrimeNG harnesses this power by providing a comprehensive set of CSS variables that define colors, making it easier for you to customize the application's theme.

PrimeNG's color system is built around a set of predefined CSS variables. These variables represent a wide range of colors, from primary and secondary colors to various shades and tints. By leveraging these variables, you can ensure consistent color usage across your application and easily tweak the appearance as needed.

For instance, PrimeNG defines primary colors such as `--primary-color` and primary color text such as `--primary-color-text`. These are just the tip of the iceberg, as there are variables for text color, surface, and highlight, among others.

> **Note**
>
> For a full list of color variables, please visit `https://primeng.org/colors`.

Using these color variables is straightforward. Instead of hardcoding color values in your styles, you reference the PrimeNG color variables. This not only ensures consistency but also makes future color changes a breeze.

Here's a simple example:

```
.my-custom-button {
    background-color: var(--primary-color);
    border-radius: var(--border-radius);
}
```

In the code, the button's background color is set to PrimeNG's `--primary-color`, and the border radius is set to `--border-radius`. If you ever decide to change the primary color or border radius in the future, the button's appearance will automatically update, without any need to modify the `.my-custom-button` styles.

Customizing PrimeNG color variables

One of the major advantages of CSS variables is the ability to be overridden. If you wish to customize the default colors provided by PrimeNG, you can easily do so by redefining the variables in your styles.

For example, here's how to change the primary color:

```
// styles.scss

:root {
    --primary-color: #3498db; /* Your desired color */
}
```

By setting this at the root level, you effectively change the primary color across all components and elements that use the `--primary-color` variable.

Other tips and tricks

In this section, we will get to know some valuable tips, techniques, and best practices that will be helpful when working with theming:

- **Global styles**: If you want to apply styles globally across your application, you can define them in a global CSS file and include it in your application. This way, you can customize common elements such as typography, colors, and layout.
- **Stay up to date**: PrimeNG is actively developed. New versions might introduce changes. Always check the documentation and update your styles accordingly.

- **Use a base theme**: When starting a new project, consider using one of the pre-built themes as a base. It gives you a solid foundation, and you can then override specific parts as needed. You can check the list of built-in themes at `https://primeng.org/theming#builtinthemes`.
- **Component variables**: Some PrimeNG components expose SCSS variables, allowing for easier customization. Dive into the source code or documentation to discover these gems. Here is an example of button styling from `theme-base`:

```
.p-button {
  ...
    &:enabled:hover {
        ...
        border-color: $buttonHoverBorderColor;
    }
}
```

  From the code, you can see that the button utilizes the `$buttonHoverBorderColor` variable on the border color when the button is hovered over. This variable is declared in `variables/_button.scss` under `mytheme`:

```
/// Border color of a button in hover state
/// @group button
$buttonHoverBorderColor: $primaryDarkColor;
```

  When updating the value of `$buttonHoverBorderColor` under `_button.scss`, it will reflect in all components that utilize this `$buttonHoverBorderColor` variable, such as the button hover state.

Note

The provided button component variable examples are just a small subset. For more information and comprehensive code associated with the buttons, you can check out the following links: `https://github.com/primefaces/primeng-sass-theme/blob/main/theme-base/components/button/_button.scss` and `https://github.com/primefaces/primeng-sass-theme/blob/main/themes/mytheme/variables/_button.scss`.

- **Test across browsers**: Always test your styles across different browsers to ensure consistency. Some components might have browser-specific styles.

In short, the most important thing is to stay updated with the changes from the PrimeNG and Angular teams. It will be beneficial if we can utilize the latest practices to improve our applications. Now, let's summarize the key points we covered throughout this chapter and reflect on the essential takeaways.

Summary

Throughout this chapter, we delved deep into the world of PrimeNG theming, embarking on a journey that took us from understanding the basics of PrimeNG theming to mastering the art of customizing components to fit our unique needs.

We began by introducing the concept of PrimeNG theming, emphasizing its significance in creating cohesive and visually appealing Angular applications. By leveraging PrimeNG's theming capabilities, we can ensure a consistent look and feel across our applications, enhancing the user experience.

We then explored the vast array of pre-built themes provided by PrimeNG. These themes, ranging from light to dark and everything in between, offer a quick and easy way to give our applications a professional appearance without the need for extensive customization. Venturing beyond the pre-built options, we learned how to craft our own custom themes using the visual editor. This powerful tool allows us to tweak every aspect of our application's appearance, ensuring it aligns perfectly with our brand or desired aesthetic.

As we progressed, we discussed the importance of overriding component styles and the utility classes provided by PrimeNG. These tools give us the flexibility to fine-tune our application's appearance, ensuring every detail is just right. Theming isn't just about making an application "look pretty." It's about creating a consistent, intuitive, and engaging user experience. By understanding and effectively utilizing PrimeNG's theming capabilities, we can craft applications that not only look stunning but also resonate with our target audience. This knowledge empowers us to elevate our applications, setting them apart in a crowded marketplace.

As we transition to the next chapter, we'll shift our focus to performance optimization techniques. Here, we'll uncover strategies and best practices to ensure our Angular applications run smoothly and efficiently. With the foundation of theming now solidly under our belt, we're ready to tackle the intricacies of performance and scalability.

10

Exploring Optimization Techniques for Angular Applications

In today's digital age, users expect applications to be fast and responsive. A slight delay can lead to decreased user satisfaction and can even impact business metrics. By understanding and implementing performance optimization techniques, you ensure that your applications meet user expectations, leading to better user engagement, higher retention rates, and positive user feedback. So, as we delve deeper into the world of Angular and PrimeNG, it's essential to ensure our applications are not just functional and aesthetically pleasing but also perform at their peak. This chapter is dedicated to equipping you with the tools and techniques to optimize the performance of Angular applications that utilize PrimeNG components.

Throughout this chapter, you'll delve into the core concepts of performance optimization. We'll unravel some popular techniques, such as lazy loading, `trackBy` via `*ngFor`, pure pipes, bundle optimization, and more. As we progress, you'll be introduced to the subtle nuances of change detection strategies, gaining a comprehensive understanding of their profound implications on performance. To round off, we'll immerse you in hands-on sessions with Angular's built-in performance tools. These tools stand out as true game-changers that are adept at pinpointing and rectifying performance bottlenecks.

This chapter will cover the following topics:

- Introducing Angular performance optimization
- Working with performance profiling and analysis
- Implementing lazy loading and deferring
- Working with change detection
- Optimizing data binding
- Working with code and bundle optimization

Technical requirements

This chapter contains various working code samples on Angular optimization. You can find the related source code in the `chapter-10` folder in this book's GitHub repository: `https://github.com/PacktPublishing/Next-Level-UI-Development-with-PrimeNG/tree/main/apps/chapter-10`.

Introducing Angular performance optimization

Every developer dreams of building applications that are both feature-rich and blazing fast. But as we add more features and complexity to our Angular applications, we might inadvertently introduce performance bottlenecks. That's where **Angular performance optimization** comes into play. Let's dive deep into understanding this crucial aspect of Angular development.

What is Angular performance optimization?

Angular performance optimization refers to the practice of improving the performance and responsiveness of Angular applications by applying various techniques, strategies, and best practices. It involves identifying and addressing performance bottlenecks, reducing unnecessary computations, optimizing rendering processes, and minimizing load times. By optimizing the performance of Angular applications, we can enhance user satisfaction, minimize user exits, and improve overall application success.

Major performance issues in Angular applications

Angular is a powerful framework, but like any tool, it's not immune to performance issues. Some of the common culprits are as follows:

- **Slow initial load time**: When an Angular application has a large code base, loading all the required resources can be time-consuming. Slow initial load times can significantly impact user perception and engagement.
- **Excessive change detection**: Angular's default change detection mechanism can be computationally expensive, especially in large applications with frequent updates. Inefficient change detection can lead to unnecessary rendering and performance degradation.
- **Memory leaks**: These can occur when components or services are not destroyed properly, particularly in cases where you overlook unsubscribing from an observable, leading to increased memory usage over time.
- **Complex computations**: Running complex algorithms or computations, especially in real-time scenarios such as animations, can slow down the application.

- **Inadequate lazy loading**: Loading all components and modules upfront, even if they are not immediately required, can result in increased initial load times. A lack of proper lazy loading techniques can impact the application's performance negatively.

Popular optimization techniques

To address these performance issues and optimize the performance of Angular applications, developers employ various techniques and strategies. Here are some popular optimization options:

- **Performance profiling and analysis**: Angular provides built-in performance tools and techniques that allow us to profile and analyze the performance of our applications. Tools such as Angular DevTools help identify performance bottlenecks and provide insights for optimization.
- **Lazy loading and deferring**: They are techniques that load modules or components on-demand, rather than upfront. By implementing lazy loading and deferring in our Angular applications, we can reduce the initial load time and improve overall performance.
- **Change detection strategies**: Angular provides different change detection strategies, including the `Default` strategy and the `OnPush` strategy. By choosing the appropriate change detection strategy based on the application's requirements, we can minimize unnecessary change detection and improve performance.
- **Optimized data binding**: Utilizing efficient data binding practices, such as one-time binding and `trackBy` function usage, helps minimize unnecessary updates and improve rendering performance. Optimized data binding ensures that only the necessary components are updated when changes occur.
- **Code and bundle optimization**: Minifying and bundling JavaScript and CSS files, tree shaking, and code splitting are techniques that are used to optimize the size and loading speed of Angular applications. These optimizations reduce network requests and improve overall performance.

Now that we understand Angular performance optimization and its significance, in the following sections, we will explore these techniques in detail. We will start with performance profiling and analysis.

Introducing performance profiling and analysis

Performance profiling is similar to a health checkup for your application. It involves monitoring the application's operations, understanding where it spends most of its time, and identifying potential bottlenecks. **Analysis**, on the other hand, is the subsequent step where this data is interpreted, issues are pinpointed, and strategies for performance enhancement are devised.

When is performance profiling and analysis used?

The need for performance profiling and analysis arises in the following situations:

- Parts of an application seem slower than expected
- There's a desire to ensure that new implementations don't degrade performance
- Large-scale applications need to optimize resource-intensive operations
- The goal is to offer users a seamless experience, characterized by swift load times and fluid interactions

How does performance profiling and analysis work in Angular?

Performance profiling captures data about an application's behavior during runtime. For Angular developers, a game-changer in this space is the **Angular DevTools** extension (`https://angular.io/guide/devtools`). This browser extension, tailored for Angular applications, offers insights into component tree structures and change detection cycles and even provides a dedicated performance profiling feature. It's an indispensable tool that complements the browser's native developer tools.

To utilize the Angular DevTools extension for performance profiling, follow these steps:

1. Install the Angular DevTools extension in your preferred web browser. The extension is available for Chrome and Firefox and can be installed from the respective browser's extension marketplace. In this section, I will use the Chrome browser.
2. Open the Chrome DevTools by right-clicking on an element of your Angular application and selecting **Inspect** (or by using the *Ctrl* + *Shift* + *I* keyboard shortcut).
3. In the Chrome DevTools, click on the **Angular** tab to switch to the Angular DevTools view.

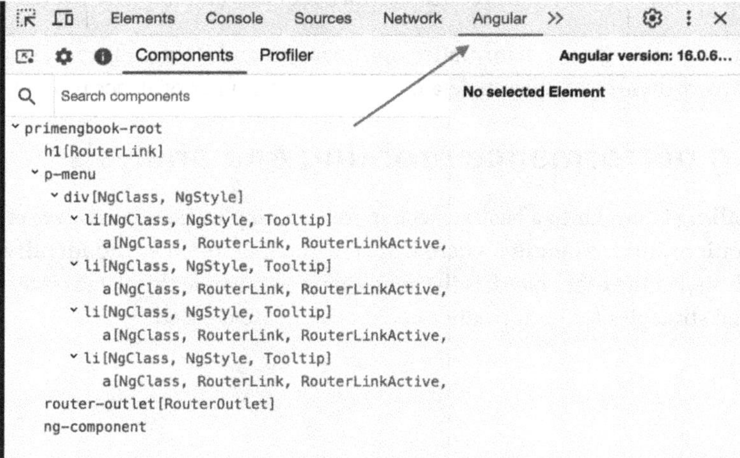

Figure 10.1 – The Angular DevTools tab

4. Within the Angular DevTools, you can enable the performance profiling feature by clicking on the red circle to collect performance data related to rendering, change detection, and other Angular-specific activities:

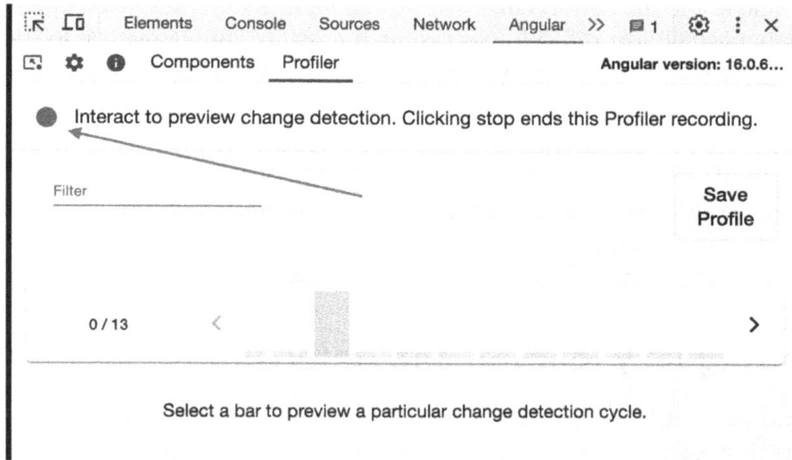

Figure 10.2 – Angular DevTools profiling

5. With performance profiling enabled, interact with your Angular application and perform typical user actions. The Angular DevTools extension will collect performance data in real time.
6. Use Angular DevTools to inspect various performance metrics and identify areas of concern. For example, you can analyze the time spent on rendering individual components, the number of change detection cycles, or the duration of network requests.
7. Based on the collected performance data, identify components, directives, or services that contribute to performance bottlenecks. Look for areas where rendering or change detection takes a significant amount of time or where unnecessary data fetching occurs.
8. Once performance bottlenecks have been identified, make targeted optimizations to the code, component structure, or data fetching mechanisms. This may involve optimizing change detection strategies, implementing memoization techniques, or introducing pagination or caching for data fetching.
9. After applying optimizations, reanalyze the performance of your Angular application using the Angular DevTools. Validate that the optimizations have improved the identified performance bottlenecks and have not introduced new issues. Iterate on the optimization process if further improvements are necessary.

By following these steps and utilizing the Angular DevTools extension, you can gain valuable insights into Angular applications' performance characteristics. This enables you to make informed optimizations and improve the overall performance of the applications.

> **Note**
>
> It's easy to fall into the trap of over-optimization. Always weigh the benefits of optimization against the effort and complexity introduced. For example, if the component in question is relatively simple and the performance gains from implementing advanced optimization techniques are negligible or difficult to measure, it might be more practical to stick with the default change detection strategy provided by Angular. This ensures a balance between optimization and development effort, avoiding unnecessary complexity and potential trade-offs in code maintainability and readability.

Performance profiling and analysis serve as the magnifying glass, revealing the intricate details of our application's behavior and efficiency. In the next section, we'll delve into the realm of lazy loading, a technique that promises to further enhance our application's responsiveness and speed.

Implementing lazy loading and deferring

Lazy loading and **deferring** are design patterns in which content is loaded only when it's needed or requested, rather than loading everything upfront. This approach optimizes resource allocation and improves performance. Here is a detailed explanation of these concepts:

- **Lazy loading**: Lazy loading is a design pattern in which content, such as Angular routes, components, or services, is loaded dynamically and on-demand, rather than everything being uploaded upfront. By implementing lazy loading, unnecessary resources are not loaded initially, resulting in a smaller initial bundle size and improved application startup times. This approach is particularly useful in large-scale applications, where loading all resources upfront can lead to performance degradation.
- **Deferring**: Deferring is a design pattern that involves delaying the loading or execution of certain resources, such as scripts or assets, until they are needed. In the context of web development, deferring typically refers to deferring the loading of JavaScript files or other resources that are not essential for the initial rendering and functionality of a web page. By deferring the loading of non-critical resources, the web page can be rendered and displayed more quickly, improving the perceived performance and overall user experience. This technique is often employed to prioritize and optimize the loading of critical resources, allowing the page to be usable as soon as possible while non-essential resources are loaded in the background.

When are lazy loading and deferring used?

Lazy loading and deferring are especially beneficial in the following scenarios:

- **Large applications**: For applications with numerous features and components, loading everything at once can be resource-intensive. Lazy loading ensures that only the necessary components are loaded, improving the initial load time.
- **User roles and permissions**: In applications where different users have different roles and permissions, lazy loading can be used to load components based on user roles, ensuring that users only download the features they have access to.
- **Network efficiency**: For users with slower internet connections, downloading a large application can be time-consuming. Lazy loading can help by reducing the initial download size.
- **Feature-rich platforms**: In platforms where certain features are used less frequently than others, it makes sense to defer those lesser-used features to improve the initial load time of the platform, resulting in faster rendering.

Example of lazy loading

Imagine an eCommerce platform built using Angular. This platform has various routes such as **Home**, **Product Listings**, **Product Details**, **Cart**, and **User Profile**. In a traditional loading approach, when a user visits the site, they would download all these routes/components upfront, even if they're just browsing the **Home** page. For example, without lazy loading, the routing will look like this:

```
export const appRoutes: Route[] = [
  {
    path: 'home',
    component: HomeComponent,
  },
  {
    path: 'cart',
    component: CartComponent,
  },
  {
    path: 'user-profile',
    component: UserProfileComponent,
  },
]
```

In this code, the routing name is associated with a component. When the user navigates to the home page, `HomeComponent` will be displayed:

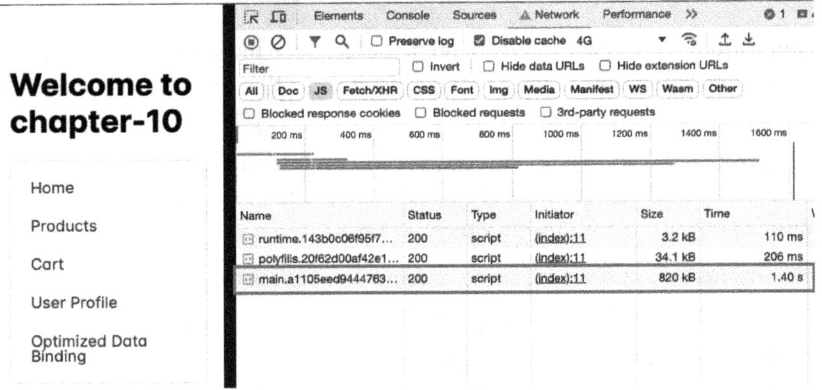

Figure 10.3 – Page without lazy loading

You'll notice that when the application is loaded, it loads all the necessary scripts from the beginning. As you can see, the main bundle size is **820 kB** and is loaded in **1.40 s**. When you navigate to other routes, their scripts are already ready-loaded. This could lead to unnecessary data consumption and a delay in page rendering, especially if the user never visits certain sections, such as **Cart** or **User Profile**, during their session. Let's see how lazy loading can improve the current implementation.

With lazy loading, the application can be structured to load the routes more intelligently:

- Initially, only the **Home** page is loaded when a user visits the site
- If a user decides to view a product, the **Products** page is loaded on the fly
- When the user wants to check their shopping cart, the **Cart** page is loaded just in time
- If they decide to view or edit their profile, the **User Profile** page is then loaded

Here's a basic example of how lazy loading can be implemented:

```
export const appRoutes: Route[] = [
  {
    path: '',
    pathMatch: 'full',
    redirectTo: 'home',
  },
  {
    path: 'home',
    loadComponent: () => import('./components/home.component'),
  },
```

```
  {
    path: 'cart',
    loadComponent: () => import('./components/cart.component'),
  },
  {
    path: 'user-profile',
    loadComponent: () => import('./components/user-profile.
component'),
  },
]
```

In the preceding code snippet, the `loadComponent` property is utilized to define functions that will load the respective component, but only when the routes are activated. This ensures that users only download the parts of the application they are actively engaging with, leading to faster load times and more efficient use of resources.

Here is the result:

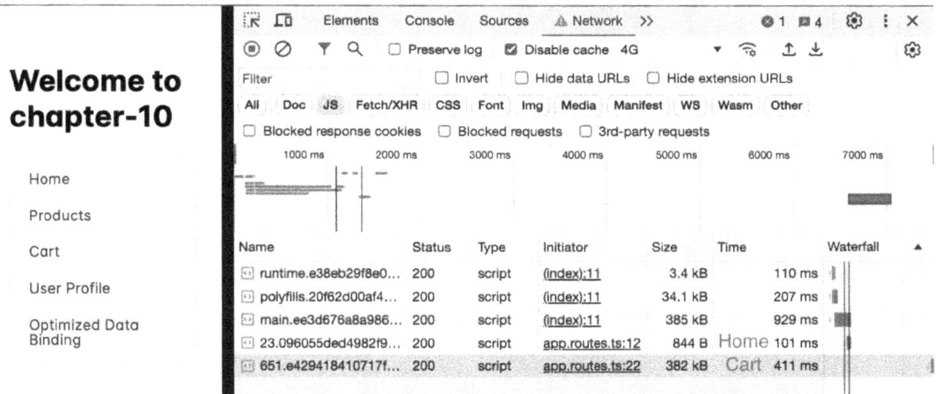

Figure 10.4 – Page with lazy loading

As you can see, the website only loads `main.js` and `HomeComponent` for the first navigation. You'll observe a significant enhancement as the main bundle size has decreased from **820 kB** to **385 kB** and **844 B**, resulting in a reduced load time of roughly 1 second compared to the previous 1.4 seconds (*Figure 10.3*). This represents a substantial performance improvement. After that, it will load `CartComponent` if you navigate to the cart route, which is initiated from `app.routes.ts`.

Example of PrimeNG deferring

PrimeNG's **Defer** directive is a tool that's designed to delay the loading of content until it enters the viewport. In essence, it's a form of lazy loading, but for various types of content. By deferring the initialization of content until it's needed, we can significantly improve the initial load time of a page.

Take the eCommerce site as an example. The site lists hundreds of products on its **Products** page. Loading all these products at once can be overwhelming and slow. With PrimeNG's `Defer`, we can optimize this.

To do so, first, ensure you've imported `DeferModule` from PrimeNG:

```
import { DeferModule } from 'primeng/defer'
```

Now, let's see how a PrimeNG `Defer` can be set up to display the product list:

```
<div class="grid gap-4" pDefer>
  <ng-template>
    <p-card
      *ngFor="let product of products"
      [header]="product.name"
      [style]="{ width: '300px' }"
    >
      <!-- product content -->
    </p-card>
  </ng-template>
</div>
```

Let's break this down:

- `<div pDefer>`: The purpose of the `pDefer` directive is to defer the rendering of its contents until later in the rendering cycle. This directive must be followed by an `ng-template` element.
- `<p-card *ngFor="let product of products">`: This element utilizes the `*ngFor` directive to iterate over an array of `products`, rendering one `<p-card>` for each product in the array.

After the implementation, the content of the product list will not appear until it becomes visible in the scroll.

PrimeNG `Defer` is also beneficial when you want to fetch product data from an API after the product list is in the viewport, which will greatly enhance the performance of your application. Here is an example:

```
// products.components.ts

<div pDefer (onLoad)="loadAnotherProducts()">
  <ng-template>
    <primengbook-product-list [products]="anotherProducts" />
  </ng-template>
</div>
```

```
...

loadAnotherProducts() {
  this.anotherProducts = this.productService.loadProducts()
}
```

Let's break down the code:

- `<div pDefer (onLoad)="loadAnotherProducts()">`: This `<div>` element is using the `pDefer` directive, which defers the loading products and rendering of its contents until the element is in the viewport.
- `loadAnotherProducts() { ... }`: This method is executed when the `(onLoad)` event is triggered. Inside the method, it calls `this.productService.loadProducts()` to load the products and assign the result to the `anotherProducts` property in the component.

> **Note**
>
> If you are working with `Table`, `Tree`, or a long list of items, you can implement `Virtual Scroller` or `Pagination` to help enhance performance and increase user experience.

Harnessing the power of lazy loading and deferring can significantly enhance the performance of our applications, ensuring that users only load what they need when they need it. In the next section, we'll delve into another crucial aspect of Angular performance: understanding and managing change detection.

Working with change detection

Change detection is the process by which Angular determines if components need to be updated in response to data changes. Whenever the data-bound properties of a component change, Angular checks if the view needs to be updated to reflect those changes. This process is automatic, but it's essential to understand its workings to optimize performance.

How does change detection work?

The change detection process in Angular follows a unidirectional flow. It starts from the root component and traverses down the component tree, checking for changes in each component. Here's a brief overview of how change detection works in Angular:

1. **Initialization**: During component initialization, Angular sets up the component's change detector and initializes the component's properties and bindings.

2. **Change detection strategy**: Angular provides two change detection strategies:

   - The `Default` strategy triggers change detection for a component whenever any of its input properties change or when an event binding is fired. Take the following component tree as an example:

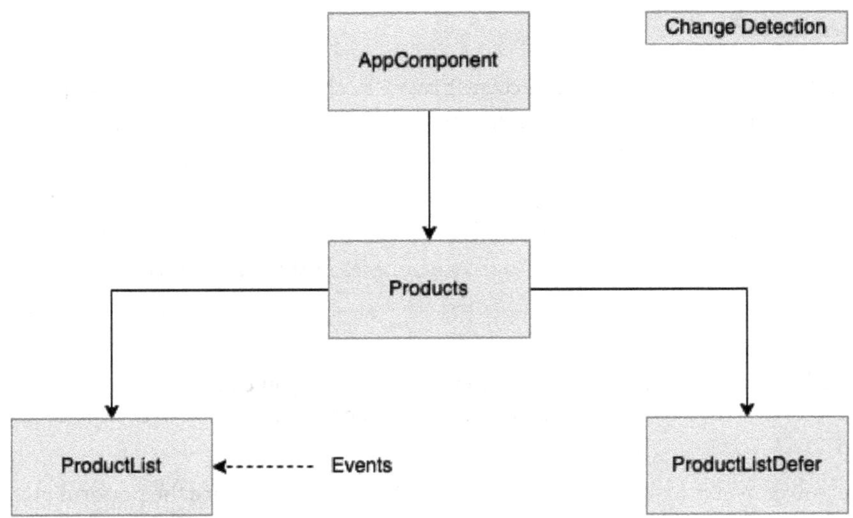

Figure 10.5 – Change detection – the default strategy

   When an event occurs on the `ProductList` component, the change detection process initiates, propagating from the root level (`AppComponent`) through all of its descendant components.

   - On the other hand, the `OnPush` strategy triggers change detection only when an input property changes or when the component receives an event from its template or a component in its view hierarchy. In the same scenario previously illustrated in *Figure 10.5*, introducing the `OnPush` change detection strategy to the `ProductListDefer` component alters the behavior of the change detection process:

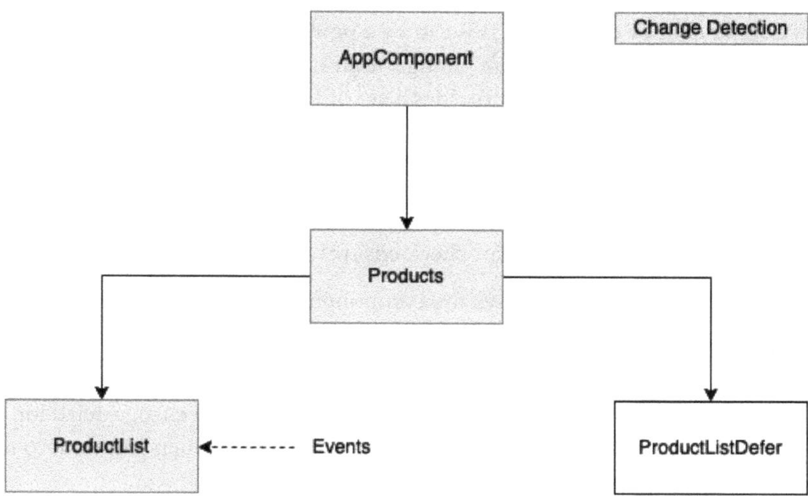

Figure 10.6 – Change detection – the OnPush strategy

Now, when an event occurs on the `ProductList` component, change detection is limited to the hierarchy from `AppComponent` to `Products` and down to `ProductList`. Notably, there is no change detection on the `ProductListDefer` component because no new reference is passed to the `ProductListDefer` component.

3. **Change detection cycle**: Angular's change detection system follows a cyclic process. In each cycle, the change detection process is performed for all components in the component tree. It involves the following steps:

   I. **Change detection check**: Angular checks the component's properties, bindings, and other inputs to detect changes. It compares the current values with the values that were stored during the previous change detection cycle.

   II. **Update view**: If a change is detected, Angular updates the component's view by updating the DOM elements associated with the changed properties or bindings.

   III. **Propagation**: If the component's view is updated, Angular propagates the changes to the child components in the component tree, triggering their change detection process recursively.

4. **Manually trigger change detection**: This can be a powerful tool in Angular when you want to explicitly update the view based on changes that might not be automatically detected. The `ChangeDetectorRef` class provides a set of functions to interact with the change detection mechanism:

   - `markForCheck()`: This function marks the component and its ancestors as needing checking during the next change detection cycle. Even if the component is not directly involved in the change, marking it for check ensures that its view will be updated.
   - `detach()`: This function detaches the component's change detector from the change detector tree. This means that the component will be skipped during the change detection process until it is reattached.
   - `detectChanges()`: Invoking `detectChanges()` triggers a change detection cycle for the component and its descendants. This is particularly useful when you want to manually check for changes in response to specific events.
   - `reattach()`: This function reverses the effect of `detach()`. It reattaches the component's change detector to the change detector tree, allowing it to participate in subsequent change detection cycles.

One of the pivotal elements in Angular's change detection mechanism is `zone.js`. This library plays a crucial role by "monkey-patching" most asynchronous operations in a browser, such as user interactions, HTTP requests, and timers.

> **Note**
>
> **Monkey patching** is a technique that allows for the modification, extension, or even suppression of the default behavior of a code segment, all without requiring any direct changes to its source code.

When these operations are completed, `zone.js` notifies Angular to run change detection. Essentially, it acts as a watchdog, keeping an eye on all asynchronous tasks. Once any of these tasks are complete, `zone.js` informs Angular to check components and update the view if necessary.

For instance, when a user clicks a button, `zone.js` detects this interaction and tells Angular that something might have changed. Angular then runs the change detection process, checking if there are any actual changes to update in the view.

> **Note**
>
> For third-party scripts, one of the techniques that can help improve the performance of your application is to load the script and run the logic outside of Angular NgZone (`https://angular.io/api/core/NgZone#runoutsideangular`).

How change detection strategies affect performance

The default change detection strategy is thorough but can be inefficient for large applications or complex component trees. Every minor data change triggers a check across all components, which can lead to lag or janky animations, especially on slower devices.

One way to improve performance is to utilize the `OnPush` strategy, in which we're telling Angular to be more selective about when it checks for changes. This can lead to a noticeable performance boost as Angular will skip checking components that we know haven't changed. However, it also means we need to be more mindful of when and how data changes to ensure our views remain up to date.

Imagine that we have two lists that display a list of products using PrimeNG components. By default, Angular checks for changes in this component and all its children every time something happens in our application. However, if our product lists don't change frequently, this can be overkill and lead to performance issues.

Let's look into the details of the default change detection strategy. Here is the sample code for the **Products** page:

```
@Component({
  standalone: true,
  imports: [CommonModule, ProductListComponent, FormsModule,
InputTextModule],
  template: `
    <h2>Products</h2>
    <div class="p-input-icon-left mb-8">
      <i class="pi pi-search"></i>
      <input
        type="text"
        pInputText
        [(ngModel)]="productName"
        (keyup)="filterProduct()"
      />
    </div>
    <primengbook-product-list [products]="filteredProducts" />
    <primengbook-product-list [products]="anotherProducts" />
  `,
})
export default class ProductsComponent {
  ...
  filterProduct() {...}
}
```

The preceding code snippet is an example of an Angular component called `ProductsComponent`. Let's break down the code and explain its functionality:

- `<input type="text" pInputText (keyup)="filterProduct()" ... />`: This defines a PrimeNG input field that executes the `filterProduct()` method whenever a key is released in the input field
- `<primengbook-product-list [products]="..." />`: This represents a custom component called `primengbook-product-list` and binds the `filteredProducts` or `anotherProducts` property of the component to the `products` input property

This is an example of a standalone component utilizing the default change detection strategy, which is set to `Default` by default. Let's observe how the application behaves in this scenario:

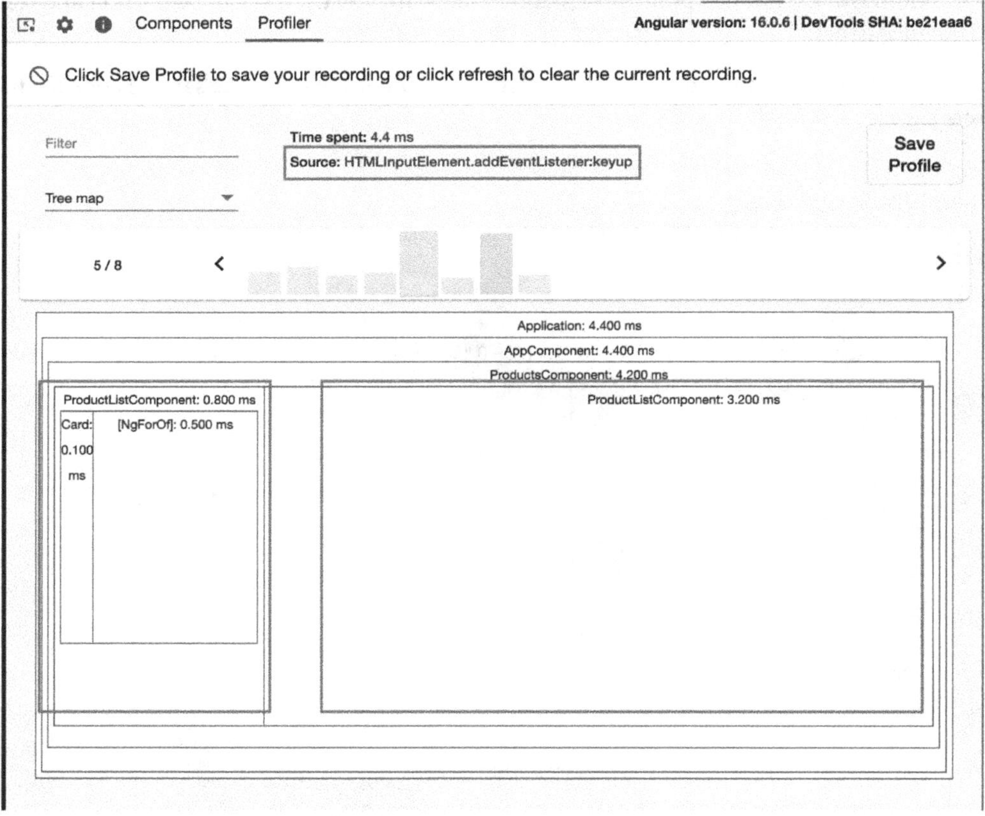

Figure 10.7 – A page using the default change detection strategy

As you can see, after we type the keyword for filtering products, it rerenders the two lists of products and the total time spent for the entire app is 4.4 ms, even though we only do the search on one product list.

The ideal situation is that when you search, it will only touch one product list. To optimize the code so that it follows this, we can set the change detection strategy to `OnPush` in `ProductListComponent`:

```
@Component({
  selector: 'primengbook-product-list',
  standalone: true,
  imports: [CommonModule, CardModule, ButtonModule],
  changeDetection: ChangeDetectionStrategy.OnPush,
  ...
})
export class ProductListComponent {
  @Input() products: Product[] = []
  ...
}
```

With this setup, Angular will only check the one `ProductListComponent` for changes when its input properties change. This means that Angular won't waste cycles checking unrelated components. Let's see how the application performs this time:

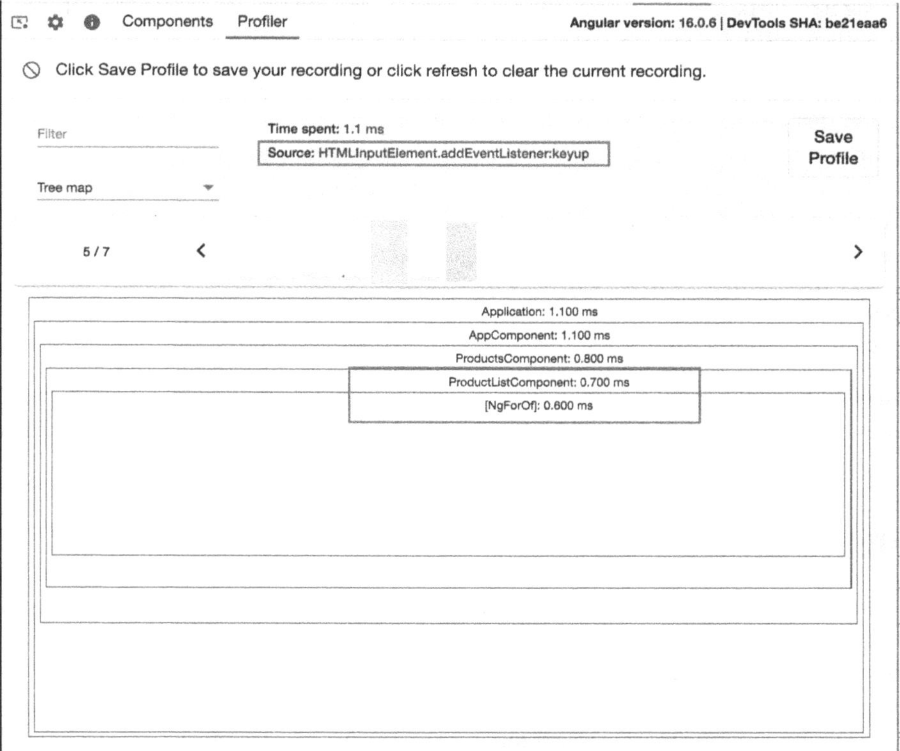

Figure 10.8 – The page with the default change detection strategy

You'll notice that when we search, our application only rerenders one product list, which reduced the time spent from 4.4 ms to 1.1 ms in total. This is a significant improvement if your application is complex.

Potential pitfalls of OnPush

While `OnPush` is powerful, it's not without its quirks. One common issue you may run into is that changes made inside the component (such as user interactions) won't trigger change detection. This can lead to situations where the UI doesn't update, even though the data has changed.

Another challenge is when working with objects and arrays. If you modify an object or array that's an input property, but the reference remains the same, Angular won't detect the change. This is where libraries such as `immutable.js` or Angular Signals come into play:

- **Immutable.js**: This offers immutable data structures such as `List`, `Map`, and `Set` that, by design, cannot be modified after creation. This immutability ensures that any change to data results in a new object reference, making it more efficient for Angular to detect differences. Furthermore, `Immutable.js` encourages the use of pure functions when modifying data, creating new instances of data structures while preserving the original data. Angular can then efficiently identify changes by comparing object references.
- **Signals** (`https://angular.io/guide/signals`): Matured in Angular v17, it's a native solution that enables Angular to optimize rendering updates more efficiently. A signal serves as a wrapper for a value and can notify interested consumers when that value changes. Signals have the flexibility to hold various types of values, ranging from basic primitives to complex data structures. Using Signals `update` also enforces an immutable approach, which is a usual practice when working with `OnPush` change detection.

I highly recommend using Signals when it's ready for production to optimize the rendering updates since it is a built-in solution from Angular instead of a third-party library.

In the realm of Angular, understanding change detection strategies is pivotal for ensuring our applications run efficiently. Next, let's delve into another crucial aspect of Angular performance: optimizing data binding to further enhance our app's responsiveness and speed.

Optimizing data binding

At its core, **data binding** in Angular is about keeping your view and component data in sync. It's the magic that allows a change in your component data to reflect instantly in your view and vice versa. However, not all data changes are equal. Some are frequent and minor, while others are rare but significant. Optimized data binding is about being selective, updating the view only when truly necessary, and doing so in the most efficient manner.

When is optimized data binding used?

Optimized data binding techniques are typically employed in the following scenarios:

- **Large datasets**: When working with large datasets, updating the entire view whenever a small portion of the data changes can be inefficient. Optimized data binding techniques help identify and update only the relevant portions of the view, minimizing unnecessary updates and improving performance.
- **Frequent updates**: In applications where data changes frequently, traditional data binding approaches can lead to excessive re-rendering and unnecessary DOM manipulations. Optimized data binding techniques help optimize data change detection and update processes to reduce overhead and improve responsiveness.
- **Complex computation**: In some cases, data binding involves computationally expensive operations, such as sorting or filtering large arrays. Optimized data binding techniques can optimize these operations by selectively updating only the affected parts of the view, rather than re-rendering the entire dataset.

A few optimized data binding techniques

Optimizing data binding in Angular revolves around a few key strategies:

- **The trackBy function**: When using `*ngFor` to loop through items, by default, Angular checks each item in the list to detect changes. By using a `trackBy` function, you can instruct Angular to track items based on their unique IDs, reducing unnecessary checks.
- **Pure pipes**: Pipes transform data in your template. A pure pipe only re-evaluates when the input changes, making it more efficient than its impure counterparts. It's crucial to bear in mind that making external requests within pipes can cause performance issues and should be avoided.
- **OnPush change detection strategy**: By utilizing the `OnPush` change detection strategy, Angular only triggers change detection for a component when its input properties change or when an event is raised within the component.
- **Immutable data structures**: Using immutable data structures can improve data binding performance. Immutable objects are not modified directly; instead, they create new instances when changes occur. This allows Angular to detect changes more efficiently and optimize rendering.

In the following sections, we will explore examples of the `trackBy` function and pure pipes.

Optimized data binding example – the trackBy function

Imagine an eCommerce platform displaying a list of products. Each product has a name, price, and rating. With thousands of products, any change in the product list, such as a price update, could potentially trigger a massive view update.

Using the `trackBy` function, we can ensure that only the affected product gets updated in the view:

```
<div *ngFor="let product of products; trackBy: trackByProductId">
  {{ product.name }} - {{ product.price | currency }}
</div>

...

trackByProductId(_: number, product: Product): number {
  return product.id
}
```

In this example, the `trackBy` function ensures that Angular only updates the view for products with actual change based on product id value.

PrimeNG also seamlessly supports the `trackBy` function. Since version 16.x, PrimeNG provides the `trackBy` property, where you can pass `trackBy` function to the data binding in the `DataView`, `OrderList`, `PickList`, and `Tree` components. Here's an example of how to use it:

```
 <p-dataView
   ...
   [trackBy]="trackByProductId"
>
   <!-- DataView content -->
</p-dataView>
```

As demonstrated in the `DataView` component, you can assign the `trackByProductId` function to the `trackBy` property. This configuration ensures that the view is only updated when the product ID value changes.

Optimized data binding example – pure pipes

Suppose you have an Angular application that displays a list of products. Each product has properties such as name, price, and quantity. The application also allows the user to calculate the product's pricing using a calculation based on the existing product quantity.

Without using pure pipes, you might have a component template that looks like this:

```
<span class="text-2xl font-semibold">
{{
  '$' + calculateTotal(product.quantity)
}}
</span>
```

In this example, the `calculateTotal` function is called for each product to calculate the total price based on the existing product quantity. Consider the scenario where the `calculateTotal` function is a custom function that performs a complex calculation. Let's look at the result:

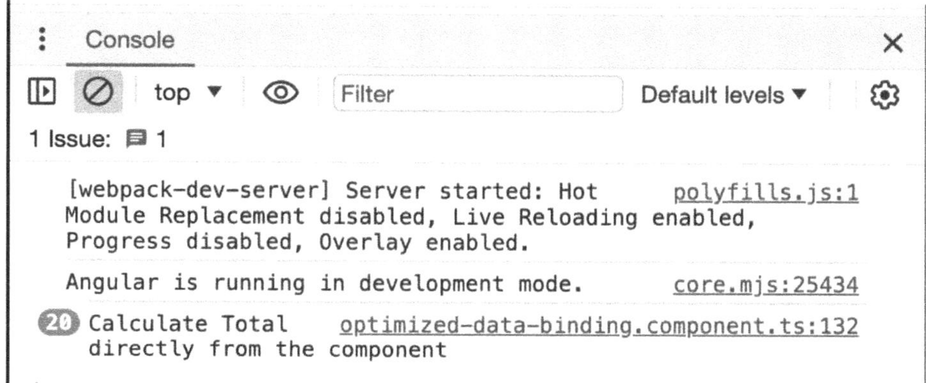

Figure 10.9 – Calculating the total price from the template

As you can see, without any optimization, every time a change is made to the product or the page, the `calculateTotal` function would be called for all products, even if only a single product's quantity has changed. This leads to unnecessary executions, even if only a single product's quantity has changed, resulting in 20 executions, as indicated in the console log. This can lead to redundant recalculations and harm performance, particularly when working with extensive datasets.

To optimize the performance using pure pipes, you can create a custom pure `pipe` called `TotalPipe` that calculates the total price for a given product:

```
import { Pipe, PipeTransform } from '@angular/core'

@Pipe({
  name: 'total',
  standalone: true,
  pure: true,
})
export class TotalPipe implements PipeTransform {
  transform(quantity: number): number {
    return calculateTotal(quantity)
  }
}
```

In this example, `TotalPipe` has been declared as a pure pipe by setting the `pure` property to `true` in the `@Pipe` decorator. This ensures that the pipe will only recalculate its output when the quantity value changes.

Now, you can use `TotalPipe` in your component template:

```
<span class="text-2xl font-semibold">
  $ {{ product.quantity | total }}
</span>
```

By using `TotalPipe`, the calculation of the total price is now delegated to the pipe. If a change is made to the quantity of a product, only the affected product's total price will be recalculated; the other products will remain unaffected:

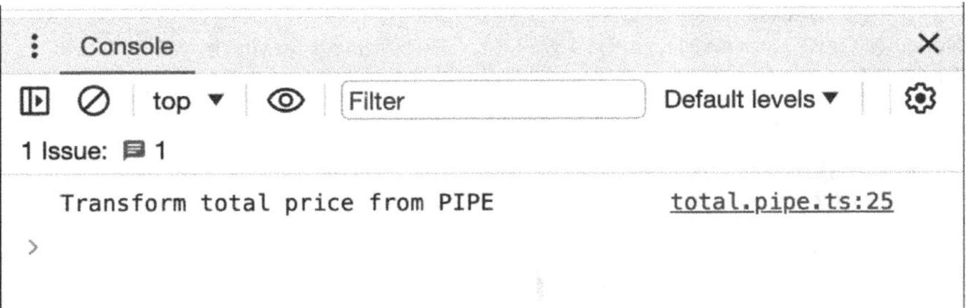

Figure 10.10 – Calculating the total price from the pipe

As we can see, when there is a change in the quality or product, it only affects one product instead of there being 20 changes (*Figure 10.9*), as in the previous approach. This optimization significantly improves performance, especially when dealing with large datasets or complex calculations.

In conclusion, optimizing data binding in web applications is essential for improving performance and reducing unnecessary re-rendering. By following best practices, you can minimize the impact of data binding on application performance. Now, let's transition to the next topic: code and bundle optimization.

Working with code and bundle optimization

Code and bundle optimization refers to the process of optimizing the code base and the resulting bundles in a web application to enhance its performance and efficiency. It involves analyzing, restructuring, and minimizing code to eliminate redundancies, reduce file sizes, and improve execution speed.

Optimizing both the code and bundle size is essential for delivering faster-loading web applications, reducing bandwidth usage, and enhancing the overall user experience. It helps ensure that the application loads quickly, responds promptly to user interactions, and performs efficiently across a range of devices and network conditions.

When is code and bundle optimization used?

Optimization isn't just a one-time task; it's an ongoing process. As soon as our application starts to grow, or when we notice performance issues, it's time to consider optimization. It's especially crucial for the following aspects:

- Large applications with extensive code bases
- Apps that rely on several third-party libraries or frameworks
- Applications targeting regions with slower internet connections
- Projects aiming for faster load times and improved user experience

Utilizing Source Map Explorer for bundle optimization

One of the most effective tools for understanding the composition of your application's bundles is **Source Map Explorer**. It provides a visual representation of the different parts of your bundle, making it easier to identify large chunks of code or unnecessary libraries that might be affecting your app's performance.

Before diving into its usage, let's set up Source Map Explorer:

1. You can add Source Map Explorer to your project by using `npm` or `yarn`. Run the following command in your project directory:

   ```
   npm install source-map-explorer --save-dev
   OR
   yarn add -D source-map-explorer
   ```

2. To generate source maps during the build process, you need to adjust the Angular build configuration. In your `angular.json` file, locate the `sourceMap` option within your application's build configurations and set it to `true`:

   ```
   // angular.json or project.json

   "build": {
     "executor": "@angular-devkit/build-angular:browser",
     "outputs": [
       "{options.outputPath}"
     ],
     "options": {
       ...
       "sourceMap": true
     },
     ...
   }
   ```

With the tool installed and the configuration adjusted, you're ready to analyze your bundles:

1. First, you need to create a production build of your Angular application with source maps. Use the following command to do so:

```
ng build --prod
```

2. After building, navigate to the `dist` folder (or wherever your build artifacts are located). You'll find several `.js` and `.js.map` files. To analyze a specific bundle, use Source Map Explorer, like so:

```
// for single file
npx source-map-explorer dist/{your-project-path}/
main.2e8930c4811df7ab.js

// for all files
npx source-map-explorer dist/{your-project-path}/**/*.js
```

3. Source Map Explorer will present a visual treemap of your bundle. Each section represents a portion of your code or an imported library. The size and position of each section correlate with its size in the bundle. Hovering over a section will display more detailed information. Here is an example:

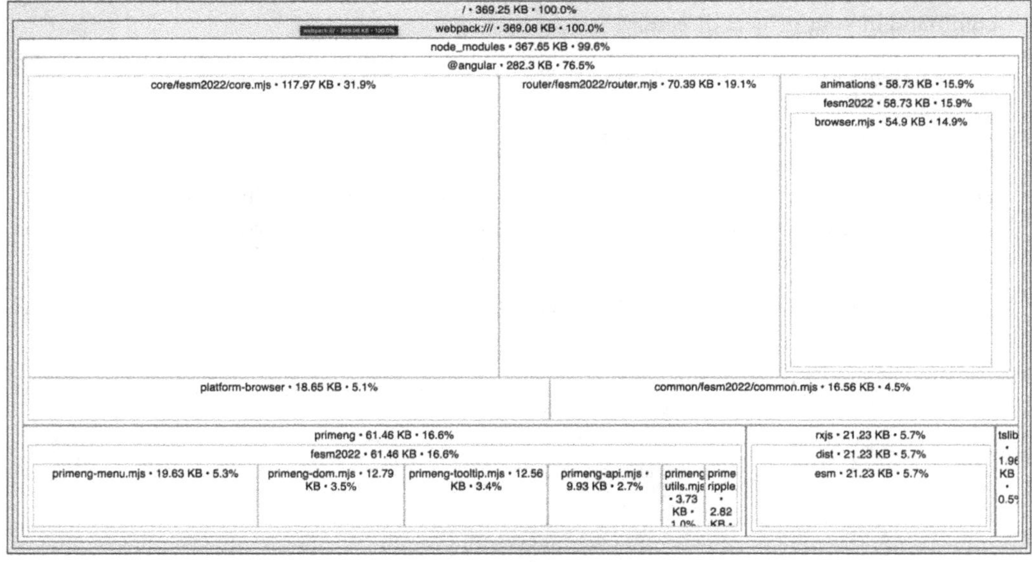

Figure 10.11 – Visual map of the bundle

The treemap's visual representation makes it easier to spot large libraries or chunks of code. If you see a library that you don't recognize or seems disproportionately large, it might be worth investigating further. You can research the unfamiliar library, analyze its dependencies, review its usage in your code base, and seek community feedback to understand its purpose and impact on performance.

> **Note**
>
> While source maps are invaluable for development and analysis, they can expose your application's code structure. Always ensure that source maps are not served in a production environment unless you want to debug on production.

In the web development world, optimizing your code and bundles is essential to achieving the best application performance. Through tools and techniques such as Source Map Explorer, we've learned how to dissect and refine our application bundles for peak efficiency. As we wrap up this chapter, let's take a moment to reflect on the key takeaways.

Summary

Throughout this chapter, we delved deep into the world of performance and optimization. We embarked on a journey to understand the intricacies of Angular performance, from the foundational concepts of lazy loading to the advanced techniques of code and bundle optimization.

At this point, you're equipped with the knowledge of how to optimize Angular applications, especially those utilizing PrimeNG components. We've seen firsthand the transformative power of efficient change detection strategies, the nuances of optimized data binding, and the importance of analyzing and refining our application bundles. These techniques are not just theoretical; they have practical implications that can drastically improve the user experience of your applications.

Why is this so crucial? In the fast-paced digital age, every millisecond counts. Users expect seamless, lightning-fast applications, and even the slightest delay can impact user retention and satisfaction. By implementing the strategies we've discussed, you're not just enhancing your application's performance; you're also ensuring that your users have the best experience possible.

As we transition to the next chapter, we'll shift our focus to building reusable components that can be easily integrated and adapted across various parts of our applications. This will empower you to write code that's not only efficient but also modular and maintainable. So, gear up for another exciting journey where we'll delve into the best practices and techniques for crafting versatile Angular components.

11

Creating Reusable and Extendable Components

In modern web development, the essence of maintainability and efficiency lies in the ability to craft components that can be effortlessly reused and extended. This not only streamlines the development process but also ensures consistency and maintainability across your applications. So, in this chapter, we will get into the concept of creating reusable and extendable components in Angular applications utilizing PrimeNG building blocks.

While PrimeNG does offer a vast selection of prebuilt reusable components, there are instances where creating custom Angular components becomes necessary. Custom components provide tailored functionality, allowing you to implement specific application requirements and integrate with existing code. They also enable UI customization, ensuring a unique visual design and user interface. Additionally, custom components allow for performance optimization and cater to domain-specific needs that may not be covered by the prebuilt components. By weighing the trade-offs between customization and development effort, you can leverage both PrimeNG's offerings and custom components to build highly adaptable and efficient Angular applications.

Throughout this chapter, we aim to equip you with the skills necessary to design components that can be effortlessly integrated into various projects, adapted as per requirements, and extended with new functionalities. This knowledge is pivotal, as it allows for rapid development without compromising on code quality.

The chapter will cover the following topics:

- Introducing reusable and extendable components
- Getting to know StyleClass
- Utilizing PrimeBlocks for creating UI blocks
- Creating reusable and extendable components
- Crafting your own components

Technical requirements

This chapter contains various working code samples on how to create reusable and extendable components. You can find the related source code in the `chapter-11` folder of the following GitHub repository: `https://github.com/PacktPublishing/Next-Level-UI-Development-with-PrimeNG/tree/main/apps/chapter-11`.

Introducing reusable and extendable components

At its core, the foundation of reusable and extendable components in Angular is built upon Angular components themselves. The only difference is that we build them in a way that allows us to encapsulate functionality and user interface elements that can be easily reused and extended in different parts of our application. By combining reusability and extensibility, we can create components that promote code reuse, maintainability, and flexibility.

In this section, we will explore why they are important and the best practices for building Angular components.

Why are reusable and extendable components important?

Reusable and extendable components offer several benefits in the development process:

- Firstly, they promote *code organization and modularity* by breaking down complex functionality into smaller, reusable units. This improves code maintainability and makes it easier to debug and test individual components.
- Secondly, reusable and extendable components *enhance development efficiency* by allowing developers to build upon existing solutions rather than starting from scratch. This saves time and effort, especially in large-scale applications where similar functionality is required in multiple places.
- Lastly, reusable and extendable components contribute to a *consistent user experience* by ensuring that common UI elements and functionalities are used consistently throughout the application. This fosters familiarity and usability for users interacting with different parts of the system.

> **Note**
> Optimizing for re-usability can be an anti-pattern—sometimes it is easier to maintain two visually identical components that are used in different contexts. As requirements continue to evolve, components can increase in complexity to cover various cases for each of the contexts.

Steps to create reusable and extendable components

Creating reusable and extendable components in Angular involves several steps. Here is a general outline of the process:

1. **Identify common patterns**: Before diving into code, thoroughly analyze your application to identify recurring UI patterns and functionalities that appear across multiple components or pages.
2. **Design with inputs and outputs**: Plan and implement your component to be flexible and adaptable, define input properties that allow data to be passed into the component from its parent, and specify output properties to emit events back to the parent when certain actions or changes occur within the component.
3. **Keep styling consistent but customizable**: Utilize Angular's encapsulated styles to ensure component styles don't affect other parts of the application. However, also consider providing customization options, such as CSS custom properties, to enable users of the component to easily override or modify its styling when needed.
4. **Write comprehensive tests**: Develop a comprehensive suite of tests to validate the functionality and behavior of your component. Include unit tests to assess individual parts of the component's logic, integration tests to verify it works correctly with other components or services, and end-to-end tests to simulate real user interactions.
5. **Document thoroughly**: Create clear and comprehensive documentation for your component. Explain its purpose, provide step-by-step instructions on how to use it, and offer guidance on extending or customizing the component if necessary. Include examples, code snippets, and any relevant API references to assist users in effectively utilizing the component in their projects.

Angular component best practices

When building components in Angular, it is important to follow best practices to maximize their effectiveness. Here are some key practices to consider:

- **Smart components and presentation components**: Angular promotes the separation of concerns through the use of smart components and presentation components. Smart components handle logic and data management, while presentation components focus on rendering the user interface and receiving input. This separation allows for better code organization and facilitates the reusability of presentation components across different smart components.
- **Component composition**: Component composition involves creating components by combining smaller, reusable components. This approach encourages the building of components with a single responsibility, making them easier to understand, test, and reuse. By composing components, you can leverage the strengths of each component and create more complex functionality by combining them.

- **Inputs and outputs**: Using inputs and outputs in components allows for easy communication between parent and child components. Inputs enable the passing of data from a parent component to a child component, while outputs facilitate the emission of events from a child component to a parent component. This interaction enables components to be more flexible and adaptable to different contexts and requirements.
- **Template and style encapsulation**: Angular provides mechanisms for encapsulating component templates and styles to prevent unintended styling conflicts and maintain component isolation. Encapsulating templates and styles within components ensure that they are self-contained and can be easily reused without affecting other parts of the application.

By creating components that encapsulate specific functionality and can be easily reused across different parts of the application, we promote code reusability and maintainability.

Now, let's transition to exploring the `StyleClass` feature, which allows us to further customize and style our components in a flexible and dynamic manner.

Getting to know StyleClass

PrimeNG **StyleClass** is a powerful feature provided by the PrimeNG library for Angular applications. It allows you to manage CSS classes declaratively, making it easier to apply dynamic styles, handle animations, and toggle classes on elements. This feature enhances the flexibility and customizability of Angular components, enabling you to create visually appealing and interactive user interfaces.

Why use PrimeNG StyleClass?

PrimeNG StyleClass is particularly useful when working with custom components in Angular. It provides a straightforward and efficient way to apply styles and manipulate classes based on various conditions or events. Using PrimeNG StyleClass, you can dynamically change the appearance and behavior of components, enhancing the user experience and adding interactivity to their applications.

One of the key advantages of PrimeNG StyleClass is its ability to handle animations during component transitions. By defining enter and leave classes, you can create smooth and visually pleasing animations when components are displayed or hidden. This can greatly enhance the overall user experience and make the application feel more polished and professional.

Another benefit of PrimeNG StyleClass is its simplicity and ease of use. You can define the target element using selector properties, which make it easier to toggle classes or add animation without creating your own custom functions.

For further reference, you can explore a comprehensive list of StyleClass properties and keywords at `https://primeng.org/styleclass`. Now, let's dive into some practical examples that demonstrate how to effectively utilize StyleClass in your projects.

Example: toggle classes

Let's consider an example where we want to create a custom Angular component that toggles dark mode based on user interaction. We can use PrimeNG StyleClass to achieve this functionality in a clean and maintainable way. Let's see the code:

```
<p-button
     pStyleClass="@next"
     toggleClass="dark-mode"
     label="Toggle Dark Mode"
/>
<p>
     <!-- CONTENT --->
</p>
```

Let's break down each element and its purpose:

- `pStyleClass="@next"`: The `pStyleClass` directive with the value `@next` indicates that the style class should be applied to the next sibling element.
- `toggleClass="dark-mode"`: This attribute specifies the class that should be toggled when the button is clicked. In this case, the class name is `dark-mode`.

Overall, the code snippet demonstrates the usage of PrimeNG's `<p-button>` component with the `pStyleClass`, `toggleClass`, and `label` attributes. It suggests that the button is used to toggle the dark mode by applying or removing the `dark-mode` class to the next element.

Here is the result:

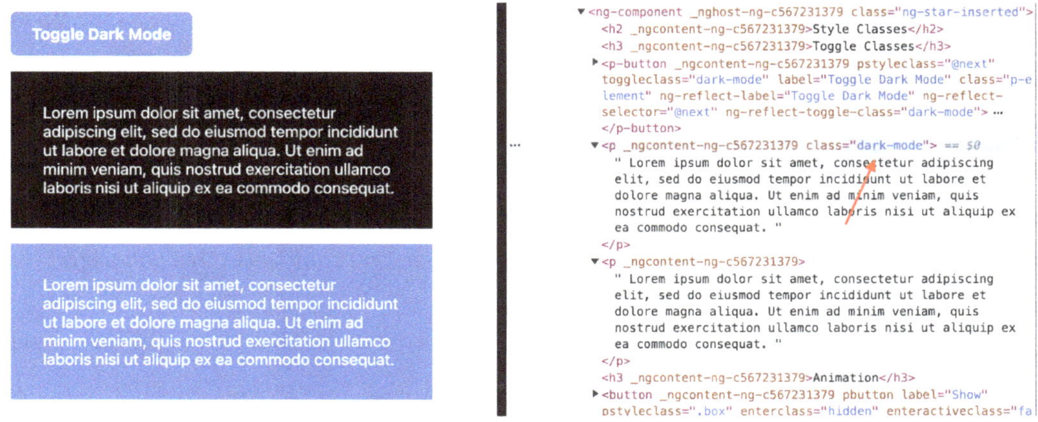

Figure 11.1 – Toggling dark mode with StyleClass

After clicking on the **Toggle Dark Mode** button, only the next paragraph gets the `dark-mode` class added, which renders the element in dark mode.

Example: animation

In addition to applying classes and styles, PrimeNG StyleClass also supports animations during component transitions. This can be achieved by specifying `enter` and `leave` classes along with their corresponding `active` and `completion` classes.

Let's consider an example where we want to create a fade-in and fade-out animation for a custom Angular component:

```
<button
    pButton
    label="Show"
    pStyleClass=".box"
    enterClass="hidden"
    enterActiveClass="fadein"
></button>
<button
    pButton
    label="Hide"
    pStyleClass=".box"
    leaveActiveClass="fadeout"
    leaveToClass="hidden"
></button>

<div class="hidden animation-duration-500 box">
    <p>
      <!-- CONTENT --->
    </p>
</div>
```

Let's break the code down:

- `<button pButton ...>`: This is the PrimeNG button component that will trigger the animation action after it is clicked.
- `pStyleClass=".box"`: This directive targets the element that has `box` in the class name.
- `enterClass="hidden"`: This defines the class `hidden` to be targeted when the button is clicked and the box content begins to appear on the screen.
- `enterActiveClass="fadein"`: This specifies the class `fadein` to be added during the `enter` animation of the associated element.

- `leaveActiveClass="fadeout"`: This defines the class `fadeout` to be added during the `leave` animation when the button is clicked; after that, the associated element starts to become hidden.
- `leaveToClass="hidden"`: This specifies the class `hidden` to be added when the `leave` animation of the box content disappears.

With a deeper understanding of `StyleClass`, we've unlocked the potential to craft visually cohesive and appealing components in our applications. Now, let's transition into utilizing PrimeBlocks for creating UI blocks, where we'll harness pre-designed blocks to further expedite our UI development process.

Utilizing PrimeBlocks for creating UI clocks

PrimeBlocks (`https://blocks.primeng.org`) is a collection of prebuilt UI blocks crafted with `PrimeFlex` developed by PrimeNG. These blocks are designed to simplify the development process by providing ready-to-use UI elements that are commonly used in web applications. PrimeBlocks offers a variety of UI blocks, including **Navbar**, **Breadcrumbs**, **Tabs**, **Footer**, **Notification**, **Dialog**, and more. These UI blocks are highly customizable and can be easily integrated into your Angular projects:

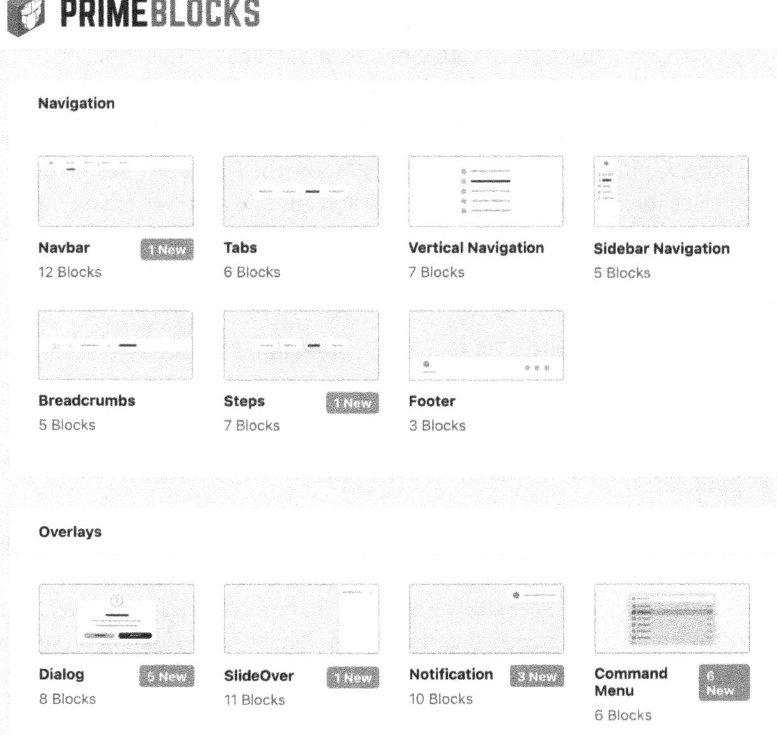

Figure 11.2 – PrimeBlocks options

PrimeBlocks offers both a free version and paid versions with distinct features. The free version provides a limited selection of options to choose from, while the paid versions offer an extensive collection of over 400 ready-to-use UI blocks. In addition to the UI blocks' codes, the paid versions also include valuable extras such as Figma files, lifetime support, and unlimited updates.

> **Note**
>
> It's important to keep in mind that the paid version has licensing restrictions, prohibiting the use of these blocks in open source projects where the code is publicly accessible.

In the next section, we'll explore the pros and cons of utilizing PrimeBlocks to provide a comprehensive overview of its benefits and considerations.

Advantages and Disadvantages of using PrimeBlocks

When compared to custom Angular components, PrimeBlocks offers several advantages:

- **Rapid prototyping**: PrimeBlocks is an excellent choice for prototyping or creating proof-of-concept applications. The library offers a wide variety of components that can be easily assembled to create a functional UI. By leveraging PrimeBlocks, you can quickly create interactive prototypes to gather feedback from stakeholders or validate design concepts. This allows for faster iteration and refinement of the application's user interface.
- **Consistent user experience**: PrimeBlocks provides a consistent and professional user experience throughout your application. The components follow the PrimeNG design language, which ensures a cohesive look and feel across different screens and sections of your application. This consistency enhances the user experience by reducing cognitive load and improving usability. By using PrimeBlocks, you can create a visually appealing and intuitive interface that aligns with industry best practices.
- **Customization and extensibility**: Despite being prebuilt UI blocks, PrimeBlocks offers a high level of customization and extensibility. Each block comes with a set of configurable options and styling classes that allow you to tailor the components to meet your specific requirements. You can easily customize colors, sizes, layouts, and behavior to match your application's branding and design guidelines.

While PrimeBlocks offers numerous benefits, there are a few potential downsides to consider. These include limited customization options, as the UI blocks are pre-designed and there's a dependency on external libraries or frameworks, licensing restrictions that prohibit usage in open source projects, and a potential learning curve in implementing the blocks effectively. Evaluating these considerations against your project requirements will help determine if PrimeBlocks is the right fit for your needs.

How to use PrimeBlocks

Since PrimeBlocks offers UI blocks that contain HTML elements, you can just copy and paste them into your Angular components. In this section, I will showcase examples utilizing the PrimeBlocks Free version. Here are just some of the blocks available:

- **Hero block**: A Hero block is a prominent section at the top of a webpage, typically featuring a captivating image or video along with a concise headline and call-to-action. It serves to grab the visitors' attention and create a visually impactful introduction to the content.
- **Feature block**: The Feature block highlights the key features or functionalities of a product, service, or website. It typically presents these features in a visually appealing manner, making it easier for users to understand and evaluate the offering.
- **Pricing block**: The Pricing block is commonly used on websites offering products or services with different pricing tiers or plans. It displays the various pricing options, including features and benefits associated with each plan, allowing users to compare and select the most suitable option.
- **Call to Action block**: The **Call to Action** (**CTA**) block is designed to prompt users to take a specific action, such as signing up, making a purchase, or subscribing to a newsletter. It usually includes a persuasive message along with a prominent button or link to encourage user engagement.
- **Banner block**: A Banner block is a horizontal section typically placed at the top or bottom of a webpage. It often contains important announcements or promotional content to capture users' attention and convey essential information.

Let's see how to use them.

Getting started

Before getting started, make sure that you have the correct version of PrimeNG (v11.4.4+) and PrimeFlex (v3.2.0+). If you haven't installed PrimeFlex yet, you can type this command to install it:

```
npm install primeflex --save
```

After that, you can add the `primeflex.scss` to `styles.scss` file:

```
@import 'primeflex/primeflex.scss';
```

Picking your blocks

After the preparation is done, you can implement blocks:

1. On the PrimeBlocks website, navigate to the **Free Blocks** section (`https://blocks.primeng.org/#/free`) and choose a block that you would like to implement. For this example, I will pick **Call to Action** block, which you can see here:

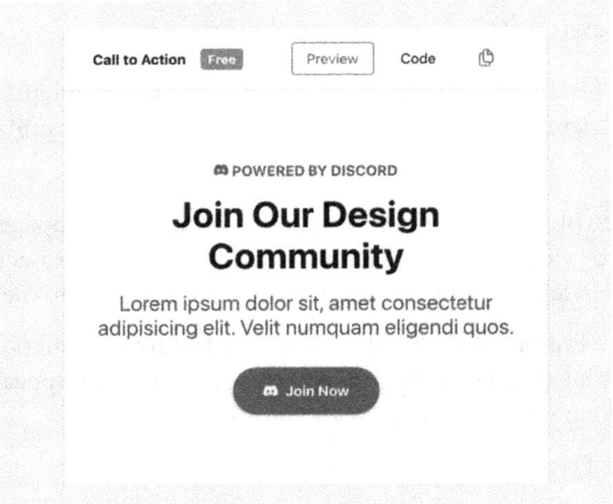

Figure 11.3 – PrimeBlocks Call to Action

2. Click on the **Code** button and you will see the code ready to use:

Figure 11.4 – PrimeBlocks Call to Action code

3. All we need to do is copy and paste the code into our application and it will work like a charm:

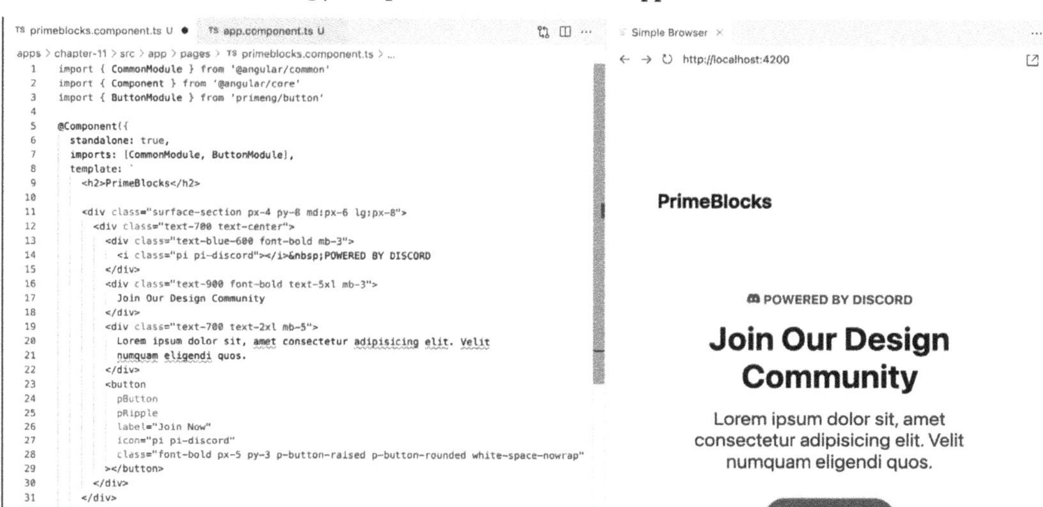

Figure 11.5 – Implement Call to Action code

Having delved into the capabilities of PrimeBlocks, we've seen how these building blocks can streamline our UI development, ensuring consistency and saving time. They offer a solid foundation for rapidly creating visually appealing and functional user interfaces. However, sometimes, our projects demand a more tailored approach. As we move on to the next section, we'll explore how to craft components that fit our unique requirements while maintaining reusability and extensibility.

Creating reusable and extendable components

As a developer, one of our key goals is to write clean, maintainable, and reusable code. In this section, we will explore the concept of creating reusable and extendable components. We'll take an example of a block from PrimeBlocks and discuss how we can make it more flexible and adaptable for different use cases.

Here is the **Stats** block from PrimeBlocks with four different stats:

Stats Free Preview Code
Orders 152 24 new since last visit
Revenue $2.100 %52+ since last week
Customers 28441 520 newly registered
Comments 152 Unread 85 responded

Figure 11.6 – PrimeBlocks Stats

After clicking the **Code** button, we can see that there are four different elements for four stats, which can be reduced and reused:

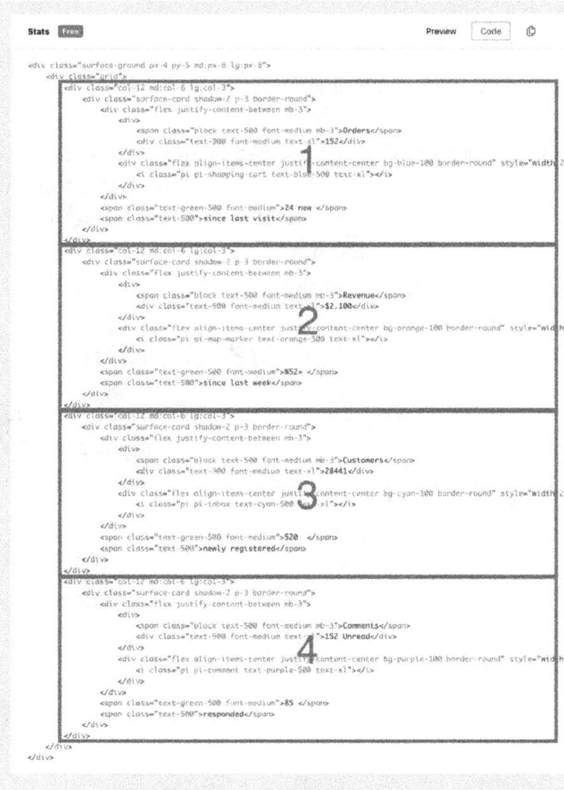

```
<div class="surface-ground px-4 py-5 md:px-6 lg:px-8">
    <div class="grid">
        <div class="col-12 md:col-6 lg:col-3">
            <div class="surface-card shadow-2 p-3 border-round">
                <div class="flex justify-content-between mb-3">
                    <div>
                        <span class="block text-500 font-medium mb-3">Orders</span>
                        <div class="text-900 font-medium text-xl">152</div>
                    </div>
                    <div class="flex align-items-center justify-content-center bg-blue-100 border-round" style="width:2.
                        <i class="pi pi-shopping-cart text-blue-500 text-xl"></i>
                    </div>
                </div>
                <span class="text-green-500 font-medium">24 new </span>
                <span class="text-500">since last visit</span>
            </div>
        </div>
        <div class="col-12 md:col-6 lg:col-3">
            <div class="surface-card shadow-2 p-3 border-round">
                <div class="flex justify-content-between mb-3">
                    <div>
                        <span class="block text-500 font-medium mb-3">Revenue</span>
                        <div class="text-900 font-medium text-xl">$2.100</div>
                    </div>
                    <div class="flex align-items-center justify-content-center bg-orange-100 border-round" style="width:
                        <i class="pi pi-map-marker text-orange-500 text-xl"></i>
                    </div>
                </div>
                <span class="text-green-500 font-medium">%52+ </span>
                <span class="text-500">since last week</span>
            </div>
        </div>
        <div class="col-12 md:col-6 lg:col-3">
            <div class="surface-card shadow-2 p-3 border-round">
                <div class="flex justify-content-between mb-3">
                    <div>
                        <span class="block text-500 font-medium mb-3">Customers</span>
                        <div class="text-900 font-medium text-xl">28441</div>
                    </div>
                    <div class="flex align-items-center justify-content-center bg-cyan-100 border-round" style="width:2.
                        <i class="pi pi-inbox text-cyan-500 text-xl"></i>
                    </div>
                </div>
                <span class="text-green-500 font-medium">520  </span>
                <span class="text-500">newly registered</span>
            </div>
        </div>
        <div class="col-12 md:col-6 lg:col-3">
            <div class="surface-card shadow-2 p-3 border-round">
                <div class="flex justify-content-between mb-3">
                    <div>
                        <span class="block text-500 font-medium mb-3">Comments</span>
                        <div class="text-900 font-medium text-xl">152 Unread</div>
                    </div>
                    <div class="flex align-items-center justify-content-center bg-purple-100 border-round" style="width:
                        <i class="pi pi-comment text-purple-500 text-xl"></i>
                    </div>
                </div>
                <span class="text-green-500 font-medium">85 </span>
                <span class="text-500">responded</span>
            </div>
        </div>
    </div>
</div>
```

Figure 11.7 – PrimeBlocks Stats code

The provided image showcases four stats in the UI (labeled from *1* to *4*), each created using nearly identical HTML blocks. This approach results in repetition and makes maintenance challenging. For instance, if you need to make styling changes, you would have to modify the code in four different places. This redundancy can lead to increased effort and potential inconsistencies when updating or refining the UI.

When examining the design and the code labeled *1*, we can look into the similarity between the elements and identify a pattern for all labels:

```
<div class="col-12 md:col-6 lg:col-3">
    <div class="surface-card shadow-2 p-3 border-round">
        <div class="flex justify-content-between mb-3">
            <div>
                <span class="block text-500 font-medium mb-3">Orders</span>
                <div class="text-900 font-medium text-xl">152</div>
            </div>
            <div class="flex align-items-center justify-content-center bg-blue-100 border-round" style="widt
                <i class="pi pi-shopping-cart text-blue-500 text-xl"></i>
            </div>
        </div>
        <span class="text-green-500 font-medium">24 new </span>
        <span class="text-500">since last visit</span>
    </div>
</div>
```

Figure 11.8 – PrimeBlocks Stats pattern

Here, we have extracted the template into reusable and customizable properties, allowing for easy modification and flexibility:

- `Orders`: `title`
- `152`: `count`
- `pi pi-shopping-cart`: `icon`
- `bg-blue-100`: `iconBackground`
- `24 new`: `newCount`
- `since last visit`: `newCountMessage`

Once this information is gathered, the pattern is extracted. We can then create a reusable presentation component like in the following example:

```
@Component({
  selector: 'primengbook-stat',
  standalone: true,
  imports: [CommonModule],
  changeDetection: ChangeDetectionStrategy.OnPush,
  template: `
    <div class="surface-card shadow-2 p-3 border-round">
      <div class="flex justify-content-between mb-3">
```

```
        <div>
          <span class="block text-500 font-medium mb-3">{{ title }}</
span>
          <div class="text-900 font-medium text-xl">{{ count }}</div>
        </div>
        <div
          class="flex align-items-center justify-content-center
{{ iconBackground }} border-round"
          style="width:2.5rem;height:2.5rem"
        >
          <i class="{{ icon }} text-blue-500 text-xl"></i>
        </div>
      </div>
      <span class="text-green-500 font-medium">{{ newCount }} </span>
      <span class="text-500">{{ newCountMessage }}</span>
    </div>
  `
})
export class StatComponent {
  @Input() title: string
  @Input() count: string
  @Input() icon: string
  @Input() iconBackground: string
  @Input() newCount: string
  @Input() newCountMessage: string
}
```

The provided code is an Angular standalone component called `StatComponent`. Let's break down its different parts and understand their purpose:

- `selector: 'primengbook-stat'`: This property specifies the HTML selector that will be used to represent this component in the template. In this case, the selector is `primengbook-stat`, meaning that this component can be used in the template as `<primengbook-stat />`.
- `changeDetection: ChangeDetectionStrategy.OnPush`: This change detection strategy tells Angular to only check for changes in the component's input properties and not perform a full change detection cycle unless triggered externally.
- `@Input() ...`: This decorator is used to mark certain properties as inputs to the component. These inputs can receive values from the component's parent or from the template where the component is used. In this case, the component has several input properties: `title`, `count`, `icon`, `iconBackground`, `newCount`, and `newCountMessage`.

After our reusable component is ready, we can use it anywhere in our application by putting it into the `imports` array:

```
@Component({
  standalone: true,
  imports: [CommonModule, StatComponent],
  template: `
    <h2>Reusable Components</h2>

    <div class="surface-ground px-4 py-5 md:px-6 lg:px-8">
      <div class="grid">
        <primengbook-stat
          *ngFor="let stat of stats"
          [title]="stat.title"
          [count]="stat.count"
          [icon]="stat.icon"
          [iconBackground]="stat.iconBackground"
          [newCount]="stat.newCount"
          [newCountMessage]="stat.newCountMessage"
          class="col-12 md:col-6 lg:col-3"
        />
      </div>
    </div>
  `,
})
export default class ReusableComponent {
  stats = [
    {
      title: 'Orders',
      count: '152',
      icon: 'pi pi-shopping-cart',
      iconBackground: 'bg-blue-100',
      newCount: '24 new',
      newCountMessage: 'since last visit',
    },
    {
      title: 'Revenue',
      count: '$2.100',
      icon: 'pi pi-shopping-cart',
      iconBackground: 'bg-orange-100',
      newCount: '%52+',
      newCountMessage: 'since last visit',
    },
  ]
}
```

The provided code is an Angular standalone component called `ReusableComponent`. Let's break down its different parts and understand their purpose:

- `imports: [CommonModule, StatComponent]`: The `imports` property is an array that specifies the modules required by this component. In this case, `CommonModule` is imported to provide common Angular directives and pipes and `StatComponent` is imported as a dependency.
- `stats = [...]`: The `stats` property is an array of objects containing the properties `title`, `count`, `icon`, `iconBackground`, `newCount`, and `newCountMessage`. These properties hold the values that will be passed to the `StatComponent` for rendering.
- `<primengbook-stat *ngFor="let stat of stats" ... />`: This code will allow you to iterate over the `stats` array and render the `primengbook-stat` elements together with their property bindings. For example, `[title]="stat.title"` binds the `title` property of the `primengbook-stat` component to the `title` property of the current `stat` object.

Here is the result:

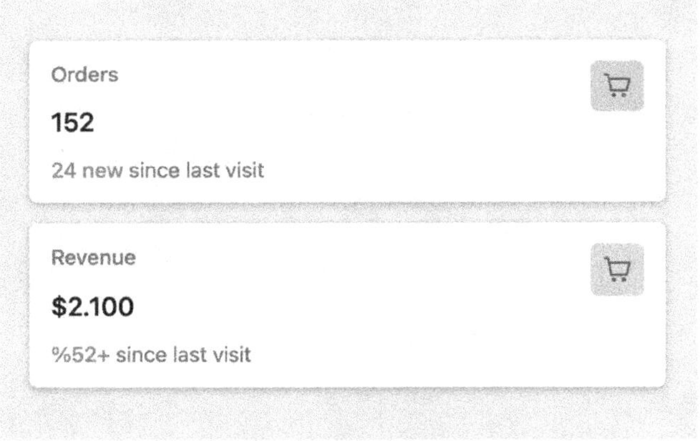

Figure 11.9 – Reusable StatComponent

Instead of having repetitive chunks of code (such as in *Figure 11.7*), using a reusable component such as `primengbook-stat` offers several benefits, including code reusability, consistent appearance and behavior, and enhanced maintainability. By encapsulating the functionality and appearance, the component can be easily reused throughout the application, reducing code duplication and promoting a modular codebase. It ensures consistent styling and interactions across different parts of the application and simplifies maintenance by allowing updates to be made in one place. Overall, it improves development efficiency and provides a better user experience.

Throughout our exploration of creating reusable and extendable components, we've uncovered the nuances of tailoring components to our specific needs without sacrificing their adaptability. This knowledge empowers us to build more efficient and maintainable Angular applications. As we transition to the next section, we'll delve deeper into how to create components from scratch, ensuring they align perfectly with our project's unique demands.

Crafting your own components with PrimeNG

Creating your own components is a fundamental aspect of modern web development. While prebuilt components can be convenient, there are situations where building custom components is necessary to meet specific requirements. In this part, we will explore the reasons behind creating custom components and learn how to leverage PrimeNG to craft your custom components.

> **Note**
>
> Always ensure that your custom components are accessible, responsive, and user friendly. Test them across different devices and browsers to ensure a consistent experience.

Why create custom components?

Pre-built components, such as those offered by PrimeNG, Bootstrap, or Material, are fantastic. They save time, ensure consistency, and often come with built-in accessibility features. However, there are scenarios where they might not be the perfect fit:

- **Unique design requirements**: Your application might have a design that doesn't align with the styles of pre-built components. In such cases, crafting your own ensures that the UI remains consistent with your design guidelines.
- **Specific functional needs**: Pre-built components offer general functionalities. If your application requires a component with very specific behavior, it's often easier to build one from scratch than to modify a pre-built one.
- **Performance**: Custom components can be optimized for your application's specific needs, potentially offering better performance in some scenarios.
- **Code maintainability and reusability**: By encapsulating specific functionality within well-defined components, you can modularize your codebase. This modular approach makes it easier to manage and maintain your application's code, as each component can be developed, tested, and updated independently.

Example: utilizing PrimeNG to create a sign-in component

PrimeNG offers a plethora of UI components and utilities that can aid in crafting your custom components. Before proceeding with the creation of the component, it is essential to take into account the following considerations:

- **Have a design ready**: It is assumed that the design for the component is already prepared using design tools such as Figma or Sketch. This ensures a clear visual reference to guide the implementation process.
- **Form and submission logic**: In this example, the focus will be on the component structure and functionality rather than the actual form submission. The implementation will omit the specific logic related to form submission.
- **Styling with PrimeFlex**: The styling of the component will be achieved using PrimeFlex utility classes. These utility classes provide a convenient and consistent way to apply styling and layout options, ensuring a cohesive and responsive design.
- **Account creation and password reset links**: This example will not include the implementation of links for creating a new account or resetting a forgotten password. The emphasis will be on the core functionality of the component itself.

Taking these points into consideration helps set the context and clarifies the scope of the component creation process. It ensures that the focus remains on the component's structure, functionality, and integration while acknowledging the design, styling, and specific features that will be excluded from this particular example.

Now let's create a sign-in component using PrimeNG. First, use Angular CLI to generate a new component:

```
ng g c sign-in
```

Then, in `sign-in.component.ts`, craft the UI:

```
@Component({
  selector: 'primengbook-sign-in',
  standalone: true,
  imports: [CommonModule, ButtonModule, InputTextModule,
CheckboxModule],
  template: `
    <div class="surface-card p-4 shadow-2 border-round w-full lg:w-6">
      ...

      <div>
        <div class="p-inputgroup mb-3">
          <span class="p-inputgroup-addon">
            <i class="pi pi-user"></i>
          </span>
```

```
          <input id="username" type="text" pInputText
placeholder="Username" />
        </div>

        <div class="p-inputgroup mb-3">
          <span class="p-inputgroup-addon">
            <i class="pi pi-lock"></i>
          </span>
          <input
            id="password"
            type="password"
            pInputText
            placeholder="Password"
          />
        </div>

        <div class="flex align-items-center justify-content-between
mb-5">
          <div class="flex align-items-center">
            <p-checkbox
              id="rememberme"
              [binary]="true"
              class="mr-2"
            ></p-checkbox>
            <label for="rememberme" class="text-900">Remember me</
label>
          </div>
          <a
            href="/forgot-password"
            class="font-medium no-underline ml-2 text-blue-500 text-
right cursor-pointer"
            >Forgot password?</a
          >
        </div>

        <button pButton type="button" label="Sign In" class="w-
full"></button>
      </div>
    </div>
  `,
})
export class SignInComponent {}
```

Let's break down the code and explain its functionality:

- `imports: [..., ButtonModule, InputTextModule, CheckboxModule]`: This line specifies the required modules that need to be imported for the components to function correctly. In this case, the component requires the `ButtonModule`, `InputTextModule`, and `CheckboxModule` modules from PrimeNG.
- `template: '...'`: This is the HTML template that consists of various elements and classes that define the visual structure and behavior of the sign-in form, such as the username field, password field, remember-me function, password reset, and the submit button.

Now, to use the sign-in component in your application, open any component/template in your directory and update its contents with the following code:

```
<primengbook-sign-in />
```

Here is the result:

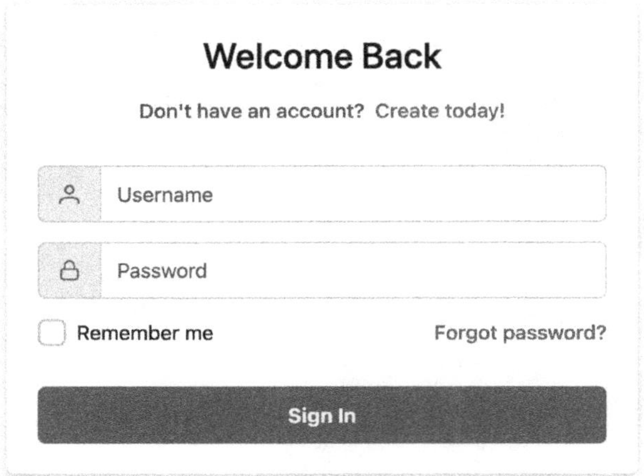

Figure 11.10 – Sign-in component

We have successfully rendered the sign-in component with the help of PrimeNG components.

In conclusion, crafting your own components empowers you to build tailored solutions that perfectly fulfill the requirements of your application. By leveraging Angular's component architecture and PrimeNG's extensive set of components and features, you can create modular, scalable, and customizable elements that enhance the user experience. Now, let's move on to the chapter summary, where we'll recap the key points covered and highlight the important takeaways from this chapter.

Summary

Throughout our journey, we delved deep into the realm of creating reusable and extendable components. The essence of this chapter was to empower you with the knowledge and techniques to craft components that can be reused across multiple parts of an application or even across different projects, all while maintaining the flexibility to extend and customize them as needed.

In this chapter, we explored why, despite the plethora of pre-built components available, there's often a need to craft our own. Whether it's to meet unique design requirements, cater to specific functional needs, or optimize performance, custom components have their place in our developer toolkit.

Moreover, using PrimeNG, we saw how to enhance our custom components. From creating a custom component to understanding the power of PrimeBlocks and StyleClass, PrimeNG proved to be an invaluable asset.

As we transition to our next topic, we'll be diving into internationalization and localization. This will be an exciting exploration into making our applications globally friendly, ensuring they cater to various languages and regional preferences. It's all about providing a seamless user experience, no matter where our users are from or what language they speak. So, gear up as we embark on this new journey, making our applications truly global.

12
Working with Internationalization and Localization

In the digital age, applications are accessed by users from every corner of the world. Catering to a global audience requires more than just translating content; it's about providing a seamless user experience that respects cultural nuances and user preferences. Ensuring that your application speaks the user's language, both literally and figuratively, can significantly enhance user satisfaction and engagement.

So, throughout this chapter, our primary objective is to equip you with the knowledge and tools necessary to make your applications universally accessible and user-friendly. We'll do this by deeply exploring the realms of internationalization and localization, guiding you through the process of making your Angular applications globally friendly with the aid of PrimeNG components. By the end of this chapter, you'll be adept at configuring language support, ensuring your application resonates with users irrespective of their geographical or cultural background.

The chapter will cover the following topics:

- Introducing internationalization and localization
- Exploring popular internationalization libraries for Angular
- Working with ngx-translate for internationalization
- Working with PrimeNG Locale

Technical requirements

This chapter contains various working code samples on internationalization and localization. You can find the related source code in the `chapter-12` folder of the following GitHub repository: `https://github.com/PacktPublishing/Next-Level-UI-Development-with-PrimeNG/tree/main/apps/chapter-12`.

Introducing internationalization and localization

In the realm of global web development, ensuring that applications cater to a worldwide audience is paramount. This is where **internationalization (i18n)** and **localization (l10n)** come into play, especially in the context of **Angular** applications.

Introducing internationalization (i18n) in Angular applications

Internationalization, often abbreviated as i18n (because there are 18 letters between the 'i' and the 'n'), is the process of designing and preparing your application to be usable in different languages. For instance, i18n requires considering language and cultural variations during product design, including the use of **Unicode** character encoding, avoiding hardcoded text, and allowing sufficient room for text expansion.

> **Note**
>
> Unicode provides universal character encoding, which means that each character is assigned a unique code point regardless of the platform, language, or application. It ensures that software can handle and display text from different languages and scripts, making applications accessible and usable across diverse linguistic and cultural contexts. You can read more at `https://unicode.org`.

There are a plethora of Angular libraries that can support this, allowing you to define translations for the content and switch between them seamlessly. We will look at these libraries shortly.

Understanding the role of localization (l10n) in creating multi-lingual experiences

Localization, abbreviated as l10n (as there are 10 letters between 'l' and 'n'), is the subsequent step after i18n. It involves adapting the internationalized application for a specific region or language by adding locale-specific translations and adjusting formats. In essence, while i18n is about making an application translatable, l10n is about doing the actual translation and adaptation.

In Angular, once the translations are provided, the application can be compiled with these translations to produce a version of the application for a specific language or locale.

Challenges and considerations for designing internationally friendly applications

Creating an application that resonates with a global audience is not without its challenges:

- **Text expansion**: Some languages might have longer translations for the same content. For example, in English, "I'm happy" can be translated as "*Tôi đang cảm thấy hạnh phúc*" in Vietnamese; this can cause the UI layout to be affected if not properly handled.
- **Right-to-left languages**: Languages, such as Arabic or Hebrew, are written from right to left, which can require significant layout adjustments. For example, the menu items that were originally on the left side need to be moved to the right side in Arabic.
- **Cultural nuances**: Colors, symbols, and images might have different cultural connotations in different regions. For example, a thumbs-up gesture is commonly used to indicate approval in Western cultures, but it can be offensive or inappropriate in some other countries.
- **Date, time, and number formats**: Different regions have different formats for displaying dates, times, and numbers. For example, in the United States, the date format is typically "mm/dd/yyyy", whereas in many European countries, it's "dd/mm/yyyy".
- **Translation management**: Managing translations efficiently can be challenging, especially in large-scale applications with numerous text strings. For instance, using translation management systems can provide features, such as translation memory, where previously translated phrases are stored and reused, reducing the effort and time required for translating repetitive content.

> **Important note**
>
> Always be aware of cultural nuances; what's acceptable or neutral in one culture might be offensive in another.

When working with Angular, it's beneficial to be aware of these challenges from the outset. By doing so, you can design your application in a way that minimizes potential issues when adding support for new languages or regions.

Exploring popular internationalization libraries for Angular

i18n is a crucial aspect of building Angular applications that cater to a global audience. Fortunately, there are several reliable i18n libraries available for Angular that simplify the process of translating and localizing applications. In this section, we will explore some popular i18n libraries for Angular, including `@angular/localize`, `ngx-translate`, `@ngneat/transloco`, and `angular-i18next`.

Here are the current stats of those libraries from NPM trends:

@angular/localize vs **@ngneat/transloco** vs **@ngx-translate/core** vs **angular-i18next**

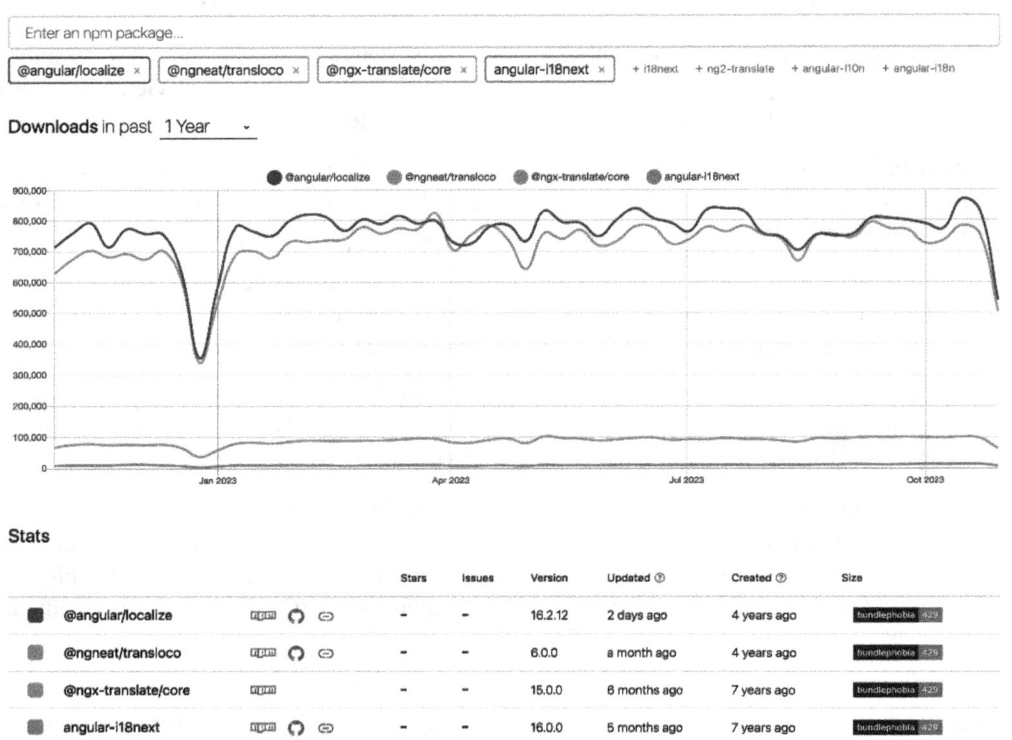

Figure 12.1 - Angular i18n library stats from NPM trends

From the trend, it is evident that `@angular/localize` and `ngx-translate` are the prominent players with approximately 500,000 downloads per day, whereas `@ngneat/transloco` receives around 100,000 downloads per day and `angular-i18next` about 13,000 downloads per day. The latest updates also have been made to `@angular/localize` and `@ngneat/transloco`.

> **Important note**
>
> Considering security-related vulnerabilities and updates is a crucial aspect when selecting a third-party library. You can assess these factors by referring to `https://snyk.io/advisor`.

Let's compare their features, discuss their pros and cons, and provide recommendations for different scenarios.

@angular/localize

@angular/localize (`https://angular.io/api/localize`) is an official Angular library introduced in **Angular 9** and is now an integral part of Angular's core, providing built-in support for i18n. `@angular/localize` leverages Angular's compiler to extract and replace translatable text in templates. Here are some key features:

- **Compile-time translation**: `@angular/localize` performs translation during the build process by statically analyzing templates. This approach allows for efficient translations and eliminates the need for runtime translation libraries.
- **Message extraction**: The Angular CLI includes a command (`ng extract-i18n`) that extracts translatable text from the application's templates and generates translation files. These files can then be translated by language experts.
- **Pluralization and gender agreement**: `@angular/localize` supports pluralization and gender agreement by providing special syntax in template expressions. This allows for accurate translations in various linguistic contexts.

There are some disadvantages, though:

- Requires a compilation step during the build process, which may increase build times for larger applications
- Has limited support for dynamic content translation and runtime language switching

In general, `@angular/localize` is suitable for projects where efficient translation during the build process is a priority and dynamic translation or runtime language switching is not a requirement.

ngx-translate

ngx-translate (`https://github.com/ngx-translate/core`) is a popular third-party i18n library for Angular, providing a flexible and feature-rich solution for translating Angular applications. Here are some notable features:

- **Runtime translation**: `ngx-translate` performs translations at runtime, allowing for dynamic language switching and on-the-fly translation updates.
- **Translation loading**: Translations can be loaded from various sources, including JSON files, APIs, and even inline definitions. This flexibility makes it easy to integrate with different translation management systems.
- **Pluralization and variable substitution**: `ngx-translate` supports pluralization and variable substitution in translations, providing rich language-specific functionalities.

Here are some disadvantages:

- Splitting translations doesn't work well, and the setup is complicated
- Loading a translation at runtime may result in the **flash of content** (**FOC**) effect, where translations are loaded and applied, potentially causing a momentary visual flash

In general, `ngx-translate` is suitable for projects that require dynamic translation updates, runtime language switching, and flexible translation loading.

@ngneat/transloco

@ngneat/transloco (`https://ngneat.github.io/transloco`) is a relatively new i18n library for Angular that aims to provide a flexible and scalable approach to translations. It offers unique features and a modern approach to i18n. Here are its key features:

- **Lazy loading**: `@ngneat/transloco` supports the lazy loading of translation files, allowing for the optimized loading of translations. This is particularly beneficial for large applications with a significant number of translatable texts.
- **Scoped translations**: `@ngneat/transloco` allows for scoping translations to specific components or modules. This feature is useful when different parts of an application require different translation sets.
- **Inline translations**: `@ngneat/transloco` provides support for in-line translations, allowing developers to define translations directly in the template. This feature simplifies the translation process and reduces the need for separate translation files.

There's just one main disadvantage:

- It is a less popular library with a smaller community compared to other established i18n libraries. Since it has a smaller community, there may be fewer available resources, tutorials, and examples to help developers get started and troubleshoot issues.

In general, `@ngneat/transloco` is suitable for projects that require the lazy loading of translations, scoped translations, and a modern approach to i18n.

angular-i18next

`angular-i18next` (`https://github.com/Romanchuk/angular-i18next`) is an integration library that combines the powerful `i18next` library with Angular. `i18next` is a widely used i18n library in the JavaScript ecosystem. Here are some notable features of `angular-i18next`:

- **Feature-rich translation library**: `angular-i18next` leverages the extensive feature set of `i18next`, including support for interpolation, pluralization, context, and much more.

- **Flexible translation sources**: Translations can be loaded from various sources, such as JSON, XHR, or even from a backend API. This flexibility makes it easy to integrate with different translation management systems.
- **l10n features**: `angular-i18next` provides additional l10n features, such as date and number formatting, that can be beneficial in certain scenarios.

Here are some disadvantages:

- Requires familiarity with both Angular and `i18next`
- It is complicated, with a huge bundle size compared to the others

In general, `angular-i18next` is suitable for projects that require extensive i18n features, flexibility in translation loading, and additional l10n functionalities.

> **Important note**
>
> It's important to note that these comparisons and recommendations are not exhaustive, and the suitability of a library may vary depending on specific project requirements and constraints. You should carefully evaluate the features, trade-offs, and community support of each library before making a decision.

In the upcoming section, we will explore i18n using the `ngx-translate` library, which is also prominently featured on the official PrimeNG website.

Working with ngx-translate for internationalization

By building upon the previous section, `ngx-translate` offers a simple and adaptable method for translating Angular applications. In this section, we will delve into the steps of integrating ngx-translate into an Angular application and provide valuable insights and strategies for maximizing productivity with ngx-translate.

Integrating ngx-translate into an Angular application

To get started with `ngx-translate`, follow these steps to integrate it into your Angular application:

1. First, install the `ngx-translate` library. Open a terminal and navigate to your Angular project's root directory. Then, run the following command to install `ngx-translate`:

   ```
   npm install @ngx-translate/core @ngx-translate/http-loader
   --save
   ```

   This command will install the `core` and `http-loader` package for initializing and loading the translation from the files using `HttpClient`.

2. In your Angular application, open the `app.config.ts` file and import the necessary modules and configurations from `ngx-translate`:

```
// translation.provider.ts
import { HttpClient } from '@angular/common/http'
import { importProvidersFrom, makeEnvironmentProviders } from '@
angular/core'
import { TranslateLoader, TranslateModule } from '@
ngx-translate/core'
import { TranslateHttpLoader } from '@ngx-translate/http-loader'

export function HttpLoaderFactory(http: HttpClient) {
  return new TranslateHttpLoader(http)
}

export const provideTranslation = () =>
  makeEnvironmentProviders([
    importProvidersFrom(
      TranslateModule.forRoot({
        defaultLanguage: 'en',
        loader: {
          provide: TranslateLoader,
          useFactory: HttpLoaderFactory,
          deps: [HttpClient],
        },
      })
    ),
  ])

// app.config.ts
export const appConfig: ApplicationConfig = {
  providers: [
    ...
    provideHttpClient(),
    provideTranslation()
  ],
}
```

The provided code is responsible for configuring and providing translation functionality using `ngx-translate` in an Angular application. Let's break it down to understand its purpose:

- `export function HttpLoaderFactory() {...}`: This code exports a factory function called `HttpLoaderFactory`. This function is used to create an instance of the `TranslateHttpLoader` class, which is responsible for loading translation files using the HTTP protocol from the `/assets/i18n/[lang].json` file (in this case, lang is en). It takes an instance of the HttpClient class as a parameter and returns a new TranslateHttpLoader instance.

- export const provideTranslation = () ...: This function returns the result of calling makeEnvironmentProviders, which is a utility function provided by Angular for generating providers. This function imports the providers from the TranslateModule.forRoot({...}) method, which is responsible for configuring the ngx-translate module at the application level with the following configuration:
  - defaultLanguage: 'en': This option indicates that English is the default language. For larger projects, it's advisable to adopt the <language>-<REGION> pattern, e.g., en-US (English (United States)) or en-CA (English (Canada)). As a company expands, there might be a requirement to accommodate similar languages with variations in different regions.
  - loader: {...}: This property sets the provide property to TranslateLoader, indicating that the TranslateLoader class is used for translation loading. The useFactory property is set to HttpLoaderFactory, which is the factory function defined earlier. The deps property specifies the dependencies required by the factory function, in this case, HttpClient.
- export const appConfig...: The provideHttpClient() and provideTranslation() providers are added to the application's dependency injection system, making HttpClient and the translation functionality available throughout the application.

3. Create translation files for each supported language. For example, create a en.json file for English translations and a vi.json file for Vietnamese translations. Place these files in an appropriate directory, such as src/assets/i18n/.

   The translation files should follow a key-value structure, where the keys represent the translation keys, and the values represent the translated text. Here is an example:

```
// en.json
{
  "greeting": "Hello!",
  "welcome": "Welcome to our application."
}

// vi.json
{
  "greeting": "Xin chào!",
  "welcome": "Chào mùng đên vói úng dung cúa chúng tôi."
}
```

4. Now, you can use the `translate` pipe to translate text in your Angular templates. Simply add the `translate` pipe to the text you want to translate, passing the translation key as an argument. Here is an example:

```
@Component({
  standalone: true,
  template: `
    <h2>Ngx Translate</h2>
    <h1>{{ 'greeting' | translate }}</h1>
    <p>{{ 'welcome' | translate }}</p>
  `,
  imports: [CommonModule, TranslateModule],
})
export default class NgxTranslateComponent {}
```

By importing the `TranslateModule`, you will have access to the `translate` pipe, which will replace the translation keys with the corresponding translated text.

Let's have a look at the result:

Ngx Translate

Hello!

Welcome to our application.

Figure 12.2 – Example of English translation using ngx-translate

From the image, we can see that the `translate` pipe is working correctly. The `greeting` key now becomes `Hello!` from the `en.json` file.

Tips and tricks for working with ngx-translate

To work effectively with `ngx-translate`, consider the following tips and tricks:

Language switching

`ngx-translate` provides a convenient method for switching between different languages at runtime. You can achieve this by using the "use" method of `TranslateService`.

Take the following `LanguageComponent` as an example:

```
import { TranslateService } from '@ngx-translate/core'
import { DropdownChangeEvent, DropdownModule } from 'primeng/dropdown'
```

```
@Component({
  standalone: true,
  selector: 'primengbook-language',
  template: `
    <p-dropdown
      [options]="languages"
      optionLabel="label"
      optionValue="value"
      (onChange)="switchLanguage($event)"
    />
  `,
  imports: [CommonModule, DropdownModule],
})
export class LanguageComponent {
  private translationService = inject(TranslateService)

  languages = [
    {
      label: 'English',
      value: 'en',
    },
    {
      label: 'Tiếng Việt',
      value: 'vi',
    },
  ]

  switchLanguage(event: DropdownChangeEvent) {
    this.translationService.use(event.value)
  }
}
```

The provided code is a shared component that represents a language selector dropdown using PrimeNG's `DropdownModule` and ngx-translate's `TranslateService`. Let's break it down more:

- `<p-dropdown ... />`: This code uses PrimeNG's `p-dropdown` component to render a dropdown list. The options property is bound to the `languages` array, which contains an array of language objects. The `optionLabel` property specifies the property to use as the `label` for each option, and the `optionValue` property specifies the property to use as the `value` for each option. The `(onChange)` event binding calls the `switchLanguage()` method when the dropdown selection changes, passing the `event` object.

- `switchLanguage(event) {...}`: This method is called when the dropdown selection changes. It receives the `event` object of type `DropdownChangeEvent` and uses `translationService` to switch the active language to the selected value from the dropdown menu.

Here is the result:

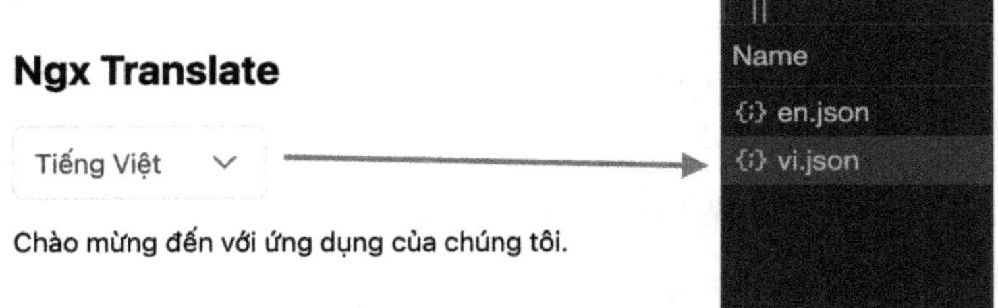

Figure 12.3 – Switch language to Vietnamese

From the image, you can see that when we switch the language from English to Vietnamese, the application dynamically loads the `vi.json` file containing the translated Vietnamese version. Consequently, the changes are promptly reflected at runtime.

Lazy loading

`ngx-translate` provides a useful feature called lazy loading, which allows translations to be loaded on-demand when they are needed. This feature is particularly beneficial for large applications with multiple languages, as it helps optimize the initial loading time by only loading translations when they are required.

To enable lazy loading in `ngx-translate`, you need to modify the translation loader configuration. Instead of loading all translation files upfront, you can set up the loader to load translations dynamically as they are requested. Here is an example of the translation files:

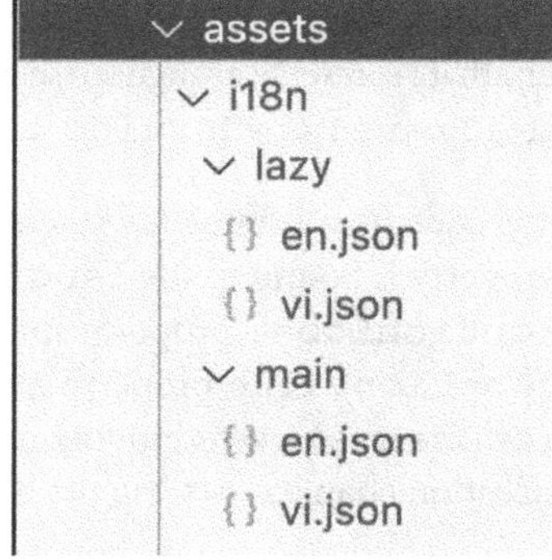

Figure 12.4 – Reorganize translation files

Based on the image, we transfer all the translation content located at the root level to the `assets/i18n/main` directory. As for the translation files for the lazy-loaded components, we can store them in a separate path, such as `assets/i18n/lazy`, which will be loaded as needed.

Here is the updated code for our main translation loader:

```
// translation.provider.ts

export function HttpLoaderFactory(http: HttpClient) {
  return new TranslateHttpLoader(http, './assets/i18n/main/', '.json')
}
```

You can see that the path for loading the translation files changed from `./assets/i18n` to `./assets/i18n/main`.

After that, we create the translation configuration for the lazy loading components:

```
// lazy-translation.provider.ts

function HttpLoaderFactory(http: HttpClient) {
  return new TranslateHttpLoader(http, './assets/i18n/lazy/', '.json')
}

export const provideLazyTranslation = () =>
  makeEnvironmentProviders([
    importProvidersFrom(
      TranslateModule.forChild({
        defaultLanguage: 'en',
        isolate: true,
        loader: {
          provide: TranslateLoader,
          useFactory: HttpLoaderFactory,
          deps: [HttpClient],
        },
      })
    ),
  ])

// app.routes.ts

export const appRoutes: Route[] = [
  ...
  {
    path: 'ngx-lazy-translate',
```

```
    loadComponent: () => import('./pages/ngx-lazy-translate.
component'),
    providers: [provideLazyTranslation()],
  },
]
```

In the code snippet, the configuration for lazy loading translation is the same as the main configuration (`translation.provider.ts`). However, there are some differences:

- `function HttpLoaderFactory(http: HttpClient) { ... }`: The function returns a new instance of `TranslateHttpLoader`, configured to load translation files from the `'./assets/i18n/lazy/'` directory with a `.json` file extension.
- `TranslateModule.forChild({..})`: This function is for the lazy load module/route. It takes an `options` object containing various configuration properties:
  - `isolate: true`: This indicates that the translations should be isolated for the lazy-loaded module, meaning they won't be shared with other modules
  - `useFactory: HttpLoaderFactory`: This specifies the translation loaders that will load the translation files from `HttpLoaderFactory` configuration
- `providers: [provideLazyTranslation()]`: This specifies an array of providers for this route. In this case, it includes the result of calling the `provideLazyTranslation()` function, which will trigger the lazy loading translations process, as explained in the previous code explanation.

Let's have a look at the result:

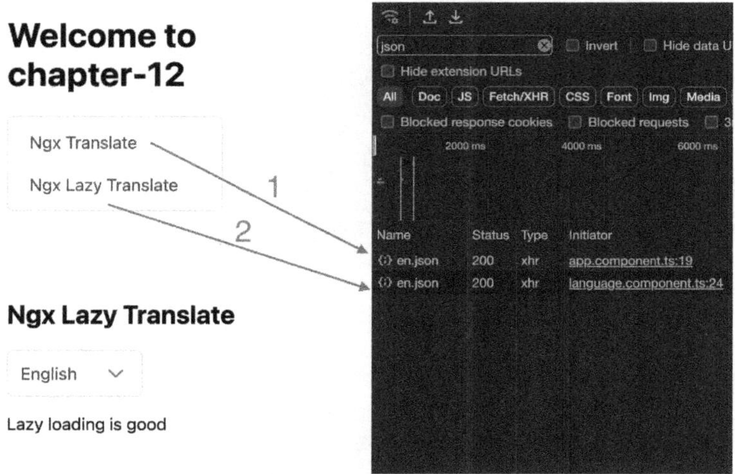

Figure 12.5 – Lazy load translation

Now, in the image, you'll observe the application's loading behavior. Upon the initial load, the translation file (labeled **1**) will be loaded. However, upon navigating to the `ngx-lazy-translate` route, the application will dynamically load the translation file specific to that route (labeled **2**) and correctly display the translated values.

Right-to-left languages

In **right-to-left** (**RTL**) languages, the layout of the user interface is mirrored in comparison to **left-to-right** (**LTR**) languages. This includes the reversal of the order of elements, such as text, buttons, and navigation. The primary goal is to align content to the right side of the screen, creating a natural flow for RTL readers.

When working with an i18n library, a common practice is to set the direction of the text after switching the language. Here is the example code:

```
private updateHtmlTag(lang: string) {
  let direction = 'ltr'

  if (['ar'].includes(lang)) {
    direction = 'rtl'
  }

  this.document.getElementsByTagName('html')[0].setAttribute('lang',
lang)
  this.document.getElementsByTagName('body')[0].setAttribute('dir',
direction)
}
```

The provided code represents a method that updates the HTML tag attributes of the document according to the specified language parameter. By default, the text direction is left-to-right (`ltr`). However, if the language is changed to Arabic (`ar`), the code will update the text direction to right-to-left (`rtl`) for the body element, and the `lang` attribute for the `html` element will also be changed to `ar`.

Let's have a look at the result:

Figure 12.6 – Right-to-left languages

From the image, you can see that after we switch the language to Arabic (`ar`), the layout is shifted from left to right to right to left.

In conclusion, by utilizing ngx-translate, you can easily handle multi-language support and dynamically switch between the different language versions of their applications. Now, let's transition to exploring **PrimeNG Locale**, another essential tool for configuring and customizing l10n settings for PrimeNG components.

Working with PrimeNG Locale

PrimeNG Locale is a feature provided by PrimeNG that enables you to configure and customize the l10n settings for its components. It allows you to define the locale-specific configurations, such as language, date format, time format, and number format. By utilizing PrimeNG Locale, you can ensure your application caters to the needs of users from different regions and cultures.

To utilize PrimeNG Locale, it is necessary to comprehend the configuration and application process within your Angular project. By default, PrimeNG only includes English translations for the locale. To localize PrimeNG components, manually updating the translations is required.

For example, the following are locale options along with their corresponding translations for the PrimeNG `Calendar` component:

src > app > components > api > TS primengconfig.ts > PrimeNGConfig > translation

```
dayNames: ['Sunday', 'Monday', 'Tuesday', 'Wednesday', 'Thursday', 'Friday',
'Saturday'],
dayNamesShort: ['Sun', 'Mon', 'Tue', 'Wed', 'Thu', 'Fri', 'Sat'],
dayNamesMin: ['Su', 'Mo', 'Tu', 'We', 'Th', 'Fr', 'Sa'],
monthNames: ['January', 'February', 'March', 'April', 'May', 'June', 'July',
'August', 'September', 'October', 'November', 'December'],
monthNamesShort: ['Jan', 'Feb', 'Mar', 'Apr', 'May', 'Jun', 'Jul', 'Aug', 'Sep',
'Oct', 'Nov', 'Dec'],
```

Figure 12.7 – Example of translation for calendar component

Based on the code, it is clear that PrimeNG incorporates these wordings for English translation within the `Calendar` component. This indicates that upon opening the `Calendar` component, you will observe the relevant text displayed:

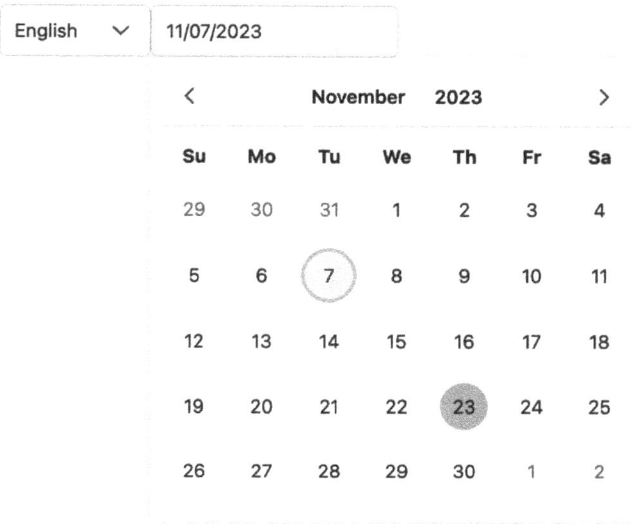

Figure 12.8 – Calendar component in English

After opening the date picker, you can observe that the month and day come from the `monthNames` and `dayNamesMin` options.

> **Important note**
>
> PrimeNG also provides a community-supported `PrimeLocale` repository (`https://github.com/primefaces/primelocale`) that you utilize or contribute to the translation content.

Now, let's take an example of how to add the translation to the PrimeNG `Calendar` component:

1. To incorporate translated versions of the locale options from PrimeNG, you have two options: either engage a translator to provide the translations or retrieve them from the `PrimeLocale` repository. As an illustration, here is the translated version (*Figure 12.7*) of the locale options in Vietnamese obtained from `PrimeLocale`:

```
{
    "vi": {
        "dayNames": ["Chủ nhật", "Thứ hai", "Thứ ba", "Thứ tu",
"Thứ năm", "Thứ sấu", "Thứ bảy"],
        "dayNamesShort": ["CN", "T2", "T3", "T4", "T5", "T6",
"T7"],
        "dayNamesMin": ["CN", "T2", "T3", "T4", "T5", "T6",
"T7"],
```

```
        "monthNames": ["Tháng Giêng", "Tháng Hai", "Tháng Ba",
"Tháng Tư", "Tháng Năm", "Tháng Sáu", "Tháng Bảy", "Tháng Tám",
"Tháng Chín", "Tháng Mười", "Tháng Mười một", "Tháng Mười hai"],
        "monthNamesShort": ["Giêng", "Hai", "Ba", "Tư", "Năm",
"Sáu", "Bảy", "Tám", "Chín", "Mười", "Mười một", "Mười hai"],
          }
    }
```

2. After that, we put it into our translation file:

```
// assets/i18n/main/vi.json

{
  "greeting": "Chào mừng đến chương 12",
  "welcome": "Chào mừng đến với ứng dụng của chúng tôi.",
  "primeng": {
    "accept": "Có",
    "reject": "Không",
    "choose": "Chọn",
    "cancel": "Hủy",
    "dayNames": [
      "Chủ nhật",
      "Thứ hai",
      "Thứ ba",
      "Thứ tư",
      "Thứ năm",
      "Thứ sáu",
      "Thứ bảy"
    ],
    ...
  }
}
```

Make sure that those translations are under the `primeng` key in the translation file.

3. Finally, when switching languages, we need to extract the translated content and set the global translation via `PrimeNGConfig`. Here is the example code:

```
// language.component.ts

private primeNgConfig = inject(PrimeNGConfig)

switchLanguage(event: DropdownChangeEvent) {
...

  this.translateService.get('primeng').subscribe((res) => {
```

```
      this.primeNgConfig.setTranslation(res)
    })
  }
```

In the code snippet, after switching the language, it fetches the translation value for the `primeng` key using the `translateService.get()` method. Once the translation value is obtained, it is passed to the `primeNgConfig.setTranslation()` method to set the translation for the `primeng` library. This ensures that PrimeNG components and features display their text and messages in the appropriate language.

Let's have a look at the result:

Figure 12.9 – Example of calendar in Vietnamese

From the image, you can see that after switching the language option from English to Vietnamese, the PrimeNG `Calendar` will also have its translations updated.

To wrap up, working with PrimeNG Locale empowers you to seamlessly handle l10n in Angular applications. By leveraging PrimeNG's features, developers can easily configure language settings, customize date and time formats, and format numbers according to locale-specific conventions. The ability to adapt applications to different regions and cultures ensures a more inclusive and user-friendly experience. As we conclude this chapter, it is valuable to pause and recap the main points covered in our summary section.

Summary

Throughout this chapter, we've embarked on a journey through the realms of i18n and l10n within the Angular framework, focusing on how PrimeNG components can be leveraged to create applications that resonate with a global audience. We've delved into the intricacies of adapting applications to various locales, ensuring that every user experiences the application in a way that feels native to them.

We've uncovered the significance of i18n and l10n, understanding that while they are closely related, they serve distinct purposes. i18n is the process of designing a software application so that it can be adapted to various languages and regions without engineering changes. l10n, on the other hand, is the process of adapting internationalized software for a specific region or language by adding locale-specific components and translating text.

We discussed the different options for i18n in Angular and showcased the ngx-translate library as a powerful tool for managing translations in Angular applications. Additionally, we explored PrimeNG Locale and its configuration process for achieving effective l10n.

Moving forward, in the next chapter, we will shift our focus to testing and debugging PrimeNG components. Testing is a critical aspect of application development as it ensures the reliability and stability of your codebase. We will explore various testing techniques and learn how to effectively debug PrimeNG components to identify and resolve issues.

13

Testing PrimeNG Components

In this chapter, we will dive into the critical aspects of testing Angular applications powered by PrimeNG components. Throughout the journey, you will learn how to effectively test your PrimeNG components, ensuring their reliability and functionality.

By understanding the principles, techniques, and tools of testing, you can enhance the quality, stability, and maintainability of your Angular applications. Throughout this chapter, you will acquire fundamental knowledge on setting up tests and efficiently testing your Angular components through various examples. Additionally, you will become acquainted with best practices and libraries that can enhance and support your testing endeavors.

In this chapter, we will cover the following topics:

- Getting started with basic Angular testing
- Writing tests for PrimeNG components
- Utilizing testing tips and tricks

Technical requirements

This chapter contains various code samples for testing. You can find the related source code in the `chapter-13` folder of the following GitHub repository: `https://github.com/PacktPublishing/Next-Level-UI-Development-with-PrimeNG/tree/main/apps/chapter-13`

Getting started with basic Angular testing

In this section, we will introduce you to the fundamentals of testing in Angular applications. Testing is an essential part of the development process that allows early bug detection, faster feedback cycles, and improved code stability. Moreover, when continuously making changes to your code over time, following testing best practices can effectively mitigate potential issues, preserve existing functionality, and ensure the delivery of high-quality software to your users.

Introduction to Angular testing fundamentals

Angular provides a robust testing framework that allows you to write tests for your components, services, and other parts of your application. Testing in Angular is based on the principles of unit testing, where individual units of code are tested in isolation. This approach helps ensure that each unit functions correctly and meets the expected requirements.

Testing in Angular involves writing test cases that simulate user interactions, verify component behavior, and assert expected outcomes. These tests help you identify and fix bugs, validate business logic, and ensure that your application works as intended.

When you first install an Angular project, it comes with **Karma** and **Jasmine** as the main testing frameworks. You can see the packages and configurations under `package.json` and `angular.json` in the following code block:

```
// package.json

"devDependencies": {
  ...
  "@types/jasmine": "~4.3.0",
  "jasmine-core": "~4.6.0",
  "karma": "~6.4.0",
  "karma-chrome-launcher": "~3.2.0",
  "karma-coverage": "~2.2.0",
  "karma-jasmine": "~5.1.0",
  "karma-jasmine-html-reporter": "~2.1.0",
}

// angular.json

"test": {
  "builder": "@angular-devkit/build-angular:karma",
  "options": {
    "polyfills": [
      "zone.js",
      "zone.js/testing"
    ],
    "tsConfig": "tsconfig.spec.json",
    ...
  }
}
```

This code snippet is part of the configuration files for setting up testing in an Angular application. Let's break this code block down:

- `devDependencies: {...}`: This contains the packages for running tests in Angular. For example, `karma-jasmine-html-reporter` generates a detailed HTML report after tests run.
- `karma`: This is a popular test runner for JavaScript applications. It allows you to execute tests in multiple browsers, capture the results, and report them. When you run `ng test`, Karma launches a development server, opens specified browsers, and executes Jasmine tests in those browsers.
- `jasmine`: This is a **behavior-driven development** (**BDD**) testing framework for JavaScript. In the context of Angular, Jasmine is often used as the testing framework for writing and running unit tests.
- `"builder": "@angular-devkit/build-angular:karma"`: This is a builder provided by the Angular DevKit, and it's used as a target for running Karma tests.

In order to run a test in Angular, you can run the `npm run test` command, which will generate the following result:

```
npm run test

> primeng-book@0.0.0 test
> ng test

✓ Browser application bundle generation complete.
15 11 2023 09:42:30.460:WARN [karma]: No captured browser, open
http://localhost:9876/
15 11 2023 09:42:30.500:INFO [karma-server]: Karma v6.4.2 server
started at http://localhost:9876/
15 11 2023 09:42:30.500:INFO [launcher]: Launching browsers Chrome
with concurrency unlimited
15 11 2023 09:42:30.503:INFO [launcher]: Starting browser Chrome
15 11 2023 09:42:33.251:INFO [Chrome 119.0.0.0 (Mac OS 10.15.7)]:
Connected on socket 3D7e73rrBAAAB with id 37006674
Chrome 119.0.0.0 (Mac OS 10.15.7): Executed 0 of 3 SUCCESS (
Chrome 119.0.0.0 (Mac OS 10.15.7): Executed 1 of 3 SUCCESS (
Chrome 119.0.0.0 (Mac OS 10.15.7): Executed 2 of 3 SUCCESS (
Chrome 119.0.0.0 (Mac OS 10.15.7): Executed 3 of 3 SUCCESS (
Chrome 119.0.0.0 (Mac OS 10.15.7): Executed 3 of 3 SUCCESS (0.139 secs
/ 0.12 secs)
TOTAL: 3 SUCCESS
```

The result of the test will also appear in the terminal and the browser:

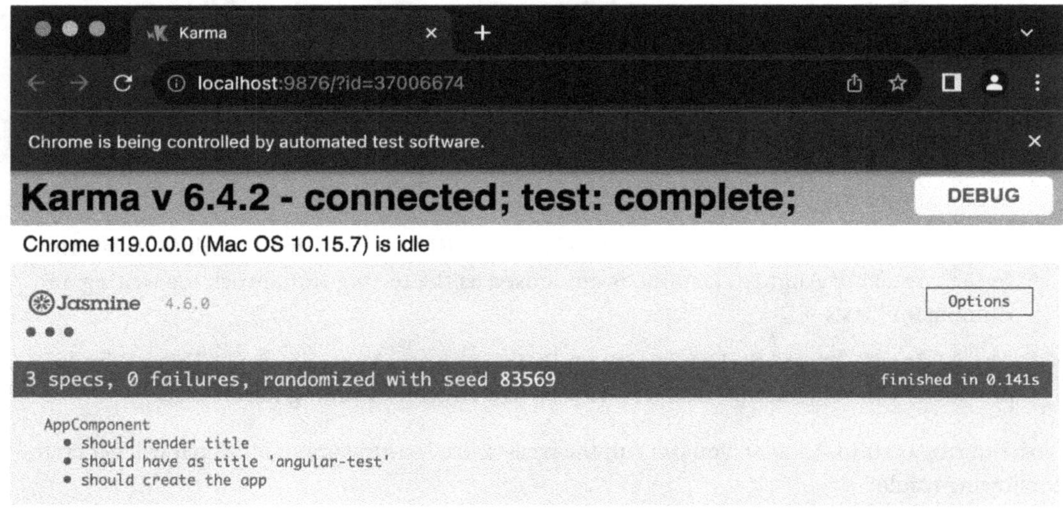

Figure 13.1 – Karma test result

As a result, when you run the command for testing, Karma will open the browser and successfully execute the tests from `AppComponent`. Additionally, you will be able to observe the executed tests and their outcomes.

Angular testing with Jest

In Angular, Karma and Jasmine have been the default testing framework choices, but with the introduction of Angular v16, **Jest** has been included as an experimental alternative for writing unit tests. The decision to adopt Jest stems from the fact that Karma, while effective, relies on a real web browser, leading to slower and heavier tests. Additionally, the use of a real browser introduces complexities in **continuous integration** (**CI**). Another factor to consider is that Karma has been deprecated (`https://github.com/karma-runner/karma`).

> **Note**
>
> CI is a development practice that involves automatically integrating code changes from multiple contributors into a shared repository. It includes automated processes such as building and testing to detect integration issues early in the development cycle. CI promotes collaboration, accelerates development, and ensures code quality by providing immediate feedback on the health of the code base. Popular CI tools include Jenkins, Travis CI, CircleCI, and GitHub Actions.

Here are some other benefits when running Jest:

- **Snapshot testing**: This simplifies the process of visually inspecting changes in the UI and helps prevent unintended regressions.
- **Fast and parallel test execution**: Jest is known for its speed and efficient test execution. It can run tests in parallel, providing quicker feedback during development.
- **Easy to integrate**: Jest is easily set up with Angular projects without requiring complex configuration.
- **Built-in code coverage reports**: Jest includes built-in support for generating code coverage reports. This allows developers to assess how much of their code is covered by tests, aiding in identifying areas that may need additional testing.
- **Mocking and spying simplified**: Jest provides a convenient API for creating mocks and spies, making it easier to isolate components or services during testing.
- **Watch mode with intelligent test re-runs**: Jest's watch mode intelligently re-runs only the tests affected by code changes, significantly speeding up the feedback loop during development.
- **Ecosystem and community**: Jest has a vibrant and active community with a growing ecosystem of plugins and extensions. This can be advantageous when looking for solutions, support, or integrations within the broader JavaScript and testing ecosystem.

So, let's change our testing framework to Jest instead. First, you need to install Jest packages by running the following command:

```
npm install jest jest-environment-jsdom --save-dev
```

This command will install `jest` and `jest-environment-jsdom`, which is an environment for Jest that simulates a browser-like environment using JSDOM.

JSDOM is a JavaScript-based library that emulates a web browser's **Document Object Model** (**DOM**) in a Node.js environment. By using `jest-environment-jsdom`, Jest tests can run in a simulated browser environment, allowing you to write and run tests that interact with the DOM, handle events, and perform other browser-related operations.

After that, let's update our testing target:

```
// angular.json

"test": {
  "builder": "@angular-devkit/build-angular:jest",
  "options": {
    "polyfills": ["zone.js", "zone.js/testing"],
    "tsConfig": "tsconfig.spec.json"
  }
}
```

By changing the builder from `@angular-devkit/build-angular:karma` to `@angular-devkit/build-angular:jest`, we now can utilize Jest for running our unit tests.

Let's have a look at the test result by running `npm run test`:

```
> primeng-book@0.0.0 test
> ng test

NOTE: The Jest builder is currently EXPERIMENTAL and not ready for
production use.

Application bundle generation complete. [1.619 seconds]
(node:22351) ExperimentalWarning: VM Modules is an experimental
feature and might change at any time
(Use `node --trace-warnings ...` to show where the warning was
created)
PASS  dist/test-out/app.component.spec.mjs
  AppComponent
    ✓ should create the app (123 ms)
    ✓ should have as title 'Welcome to chapter-13' (27 ms)
    ✓ should render title (25 ms)

Test Suites: 1 passed, 1 total
Tests:       3 passed, 3 total
Snapshots:   0 total
Time:        1.398 s
Ran all test suites.
```

You will notice that there is no difference in the process of writing unit tests and the execution speed of the tests is significantly faster.

Step-by-step guide to writing your initial Angular tests

Now that we have covered the preparation part, let's dive into writing your first set of tests. In unit testing, you can utilize the **Arrange, Act, and Assert** (**AAA**) pattern, a common methodology for structuring and organizing unit tests. Each part of AAA serves a distinct purpose:

1. **Arrange**: In this step, you set up the test environment. This involves creating instances of the components or services you want to test, providing any necessary dependencies, and configuring the initial state of the test.
2. **Act**: This is where you perform the action or trigger the behavior that you want to test. It might involve calling a method, interacting with a component, or simulating an event.

3. **Assert**: In the final step, you check the results of the action to ensure that it matches the expected outcome. This is where you make assertions about the state of your application after the action has been performed.

Using the AAA pattern helps to keep tests organized, readable, and focused on specific behaviors. It provides a clear structure for writing and understanding tests, making it easier to maintain and troubleshoot them as your code base evolves.

In the next section, let's have a look at a real Angular test.

Breaking down a simple unit test

When we first create an Angular application, the CLI helps to generate a test file for your `AppComponent`. Let's have a look at the sample test file:

```
// app.component.spec.ts

describe('AppComponent', () => {
  // Arrange
  beforeEach(async () => {
    await TestBed.configureTestingModule({
      imports: [AppComponent]
    }).compileComponents()
  })

  it('should render title', () => {
    // Arrange
    const fixture = TestBed.createComponent(AppComponent)
    const compiled = fixture.nativeElement as HTMLElement
    // Act
    fixture.detectChanges()
    // Assert
    expect(compiled.querySelector('h1')?.textContent).toContain(
      'Welcome to chapter-13'
    )
  })
})
```

The provided code snippet is an Angular test script for the `AppComponent` component. Let's break this code block down in the following list so we can understand it more:

- `describe('AppComponent', ()...`: This is a test suite that groups related tests. In this case, it groups tests for `AppComponent`.

- `beforeEach(async () => ...)`: This function is a setup function that runs before each individual test case. It is an asynchronous function that configures the testing module using `TestBed.configureTestingModule`, which is a key part of Angular's testing infrastructure provided by the `TestBed` utility. It allows you to configure the testing module by specifying the necessary dependencies, providers, and imports for the component undergoing testing. In this case, it imports `AppComponent`. This is also the Arrange step in the AAA pattern.
- `it('should render title'...`: This function is a test case that defines a specific behavior to be tested. In this case, the test case is named `should render title`. Let's break this function down based on the AAA pattern:
    - **Arrange**: We created an instance of `AppComponent` using `TestBed.createComponent`, and retrieved the compiled HTML element with `fixture.nativeElement`
    - **Act**: We triggered the ability to detect changes with `fixture.detectChanges()`
    - **Assert**: We verified that the text content of the `<h1>` element contains the string `Welcome to chapter-13`

Overall, this test verifies that `AppComponent` correctly renders a title with the expected text content. It sets up the component, triggers change detection, and asserts the expected outcome.

In short, by using basic testing techniques, we can set a solid foundation for ensuring the quality and reliability of our Angular applications. Now that we have a grasp of basic testing concepts, let's dive into a specific use case: writing tests for PrimeNG components. We will also write our unit tests while utilizing Jest as our unit testing framework.

Writing test for PrimeNG components

Testing PrimeNG components follows the same principles as testing regular Angular components. This similarity arises from the fact that PrimeNG components are essentially Angular components under the hood. In the following sections, we will explore a few of these tests to gain insights and knowledge in this area.

How PrimeNG tests its components

First, let's have a look at how PrimeNG tests its own components. Due to space limitations in this book, it is not possible to display the entire test file, as it is quite lengthy. However, we can focus on a specific section that illustrates how PrimeNG tests its components in its source code. Here is the example code of a PrimeNG `Button` component:

```
describe('Button', () => {
    let button: Button;
    let fixture: ComponentFixture<Button>;
```

```
    beforeEach(() => {
        TestBed.configureTestingModule({
            imports: [NoopAnimationsModule],
            declarations: [Button]
        });

        fixture = TestBed.createComponent(Button);
        button = fixture.componentInstance;
    });

    it('should disabled when disabled is true', () => {
        button.disabled = true;
        fixture.detectChanges();

        const buttonEl = fixture.debugElement.query(By.css('.p-
button'));
        expect(buttonEl.nativeElement.disabled).toBeTruthy();
    });

    it('should display the label and have a text only class', () => {
        button.label = 'PrimeNG';
        fixture.detectChanges();

        const buttonEl = fixture.debugElement.query(By.css('.p-
button'));
        expect(buttonEl.nativeElement.textContent).
toContain('PrimeNG');
        expect(buttonEl.nativeElement.children.length).toEqual(1);
    });

    ...
})
```

Let's break down the code in the following list so we can understand its different sections:

- `beforeEach(...)`: This function is used to set up the testing environment before each test case. In this case, it is used to configure the testing module by calling `TestBed.configureTestingModule`. The `imports` property specifies any required modules, and the `declarations` property specifies the component being tested. In this case, it imports `NoopAnimationsModule` and adds `Button` to the declarations array.

- `should disabled when disabled is true`: This test case sets the `disabled` property of the `Button` component to `true`, triggers change detection using `fixture.detectChanges()`, and then queries the DOM for the button element using `fixture.debugElement.query`. It asserts that the `disabled` property of the button element is `true` using `expect(...).toBeTruthy()`.
- `should display the label and have a text only class`: This test case sets the `label` property of the `Button` component to `PrimeNG`, triggers change detection, and queries the DOM for the `button` element. It asserts that the `button` element's text content contains `PrimeNG` using `expect(...).toContain(...)`, and it verifies that the `button` element has only one child element using `expect(...).toEqual(1)`.

The purpose of these tests is to verify the expected behavior of the `Button` component. By changing its properties and triggering change detection, the tests ensure that the component renders correctly and behaves as intended.

> **Note**
>
> In the PrimeNG source code (`https://github.com/primefaces/primeng/tree/master/src/app/components`), there are existing tests that you can study and use as references to enhance your understanding.

Creating and testing our own component

Putting the previous examples and knowledge into practice, let's create a test for the following card component:

```
template: `
  <h2>Sample Test</h2>

  <p-card
  id="card"
    header="Super Laptop PRO X"
    subheader="Best for Nomads"
    [style]="{ width: '360px' }"
  >
    <ng-template pTemplate="header">
      <img alt="Card" src="assets/placeholder.png" />
    </ng-template>
    <p>
      Lorem ipsum dolor sit amet, consectetur adipisicing elit.
      Inventore sed
      consequuntur error repudiandae numquam deserunt quisquam
      repellat libero
      asperiores earum nam nobis, culpa ratione quam perferendis
      esse,
      cupiditate neque quas!
    </p>
    <ng-template pTemplate="footer">
      <p-button label="Purchase" icon="pi pi-check"></p-button>
      <p-button
        label="Cancel"
        icon="pi pi-times"
        styleClass="p-button-secondary"
        [style]="{ 'margin-left': '.5em' }"
      ></p-button>
    </ng-template>
  </p-card>
  `,
})
export class SampleTestComponent {}
```

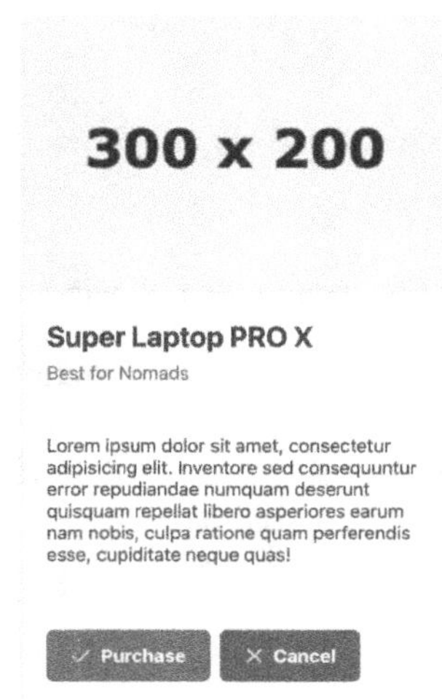

Figure 13.2 – Sample card component

By examining the screenshot, you can observe the visual representation of the card component we created. To ensure the proper functioning of this component, it is essential to write tests that validate the correct rendering of its title, subtitle, and buttons. The following code block is an example of the test code:

```
// sample-test.component.spec.ts

describe('SampleTestComponent', () => {
  let component: SampleTestComponent
  let fixture: ComponentFixture<SampleTestComponent>

  beforeEach(async () => {
    await TestBed.configureTestingModule({
      imports: [SampleTestComponent],
    }).compileComponents()

    fixture = TestBed.createComponent(SampleTestComponent)
    component = fixture.componentInstance
    fixture.detectChanges()
  })

  it('should create', () => {
    expect(component).toBeTruthy()
  })

  it('should display the product title and subtitle', () => {
    fixture.detectChanges()

    const card = fixture.debugElement.query(By.css('p-card'))
    expect(card.nativeElement.textContent).toContain('Super Laptop PRO
X')
    expect(card.nativeElement.textContent).toContain('Best for
Nomads')
  })

  it('should have a footer', () => {
    fixture.detectChanges()

    const footerCard = fixture.debugElement.query(By.css('.p-card-
footer'))
    const ctaButtons = fixture.debugElement.queryAll(By.css('.p-
button'))

    expect(footerCard).toBeTruthy()
    expect(ctaButtons).toBeTruthy()
```

```
    expect(ctaButtons.length).toEqual(2)
  })
})
```

Let's break down the code in the following list so that we can understand its different sections:

- `should create`: The test case checks whether the `component` instance is `Truthy`, meaning that it has been successfully created.
- `should display the product title and subtitle`: This test case triggers change detection and then queries the DOM for the `p-card` element using `fixture.debugElement.query`. It asserts that the content of the `p-card` element contains the expected product `title` value and `subtitle` value using `expect(...).toContain(...)`.
- `should have a footer`: This test case triggers change detection, then queries the DOM for the footer element with the `.p-card-footer` class and all the buttons with the `.p-button` class. It asserts that both the footer element and the buttons are `Truthy` using `expect(...).toBeTruthy()`, and it also verifies that there are two buttons present by checking the length of the `ctaButtons` array using `expect(...).toEqual(2)`.

After running the test, you can see that it passes successfully:

```
PASS   chapter-13  apps/chapter-13/src/app/pages/sample-test/sample-
test.component.spec.ts

SampleTestComponent

  ✓ should create (95 ms)
  ✓ should display the product title and subtitle (23 ms)
  ✓ should have a footer (17 ms)

Test Suites: 1 passed, 1 total
Tests:       3 passed, 3 total
Snapshots:   0 total
Time:        6.411 s
```

In conclusion, writing tests for PrimeNG components equips you with the ability to ensure the functionality and reliability of these components within your Angular applications. Now that we have explored the process of writing tests for PrimeNG components, let's further enhance our testing practices by utilizing helpful tips and tricks to improve the effectiveness and efficiency of our tests.

Utilizing testing tips and tricks

Even though testing enables us to ensure the correctness and stability of our code, writing effective tests can sometimes be challenging. In this section, we will explore various tips and tricks that can help you improve your testing practices, using the following editable table as an example:

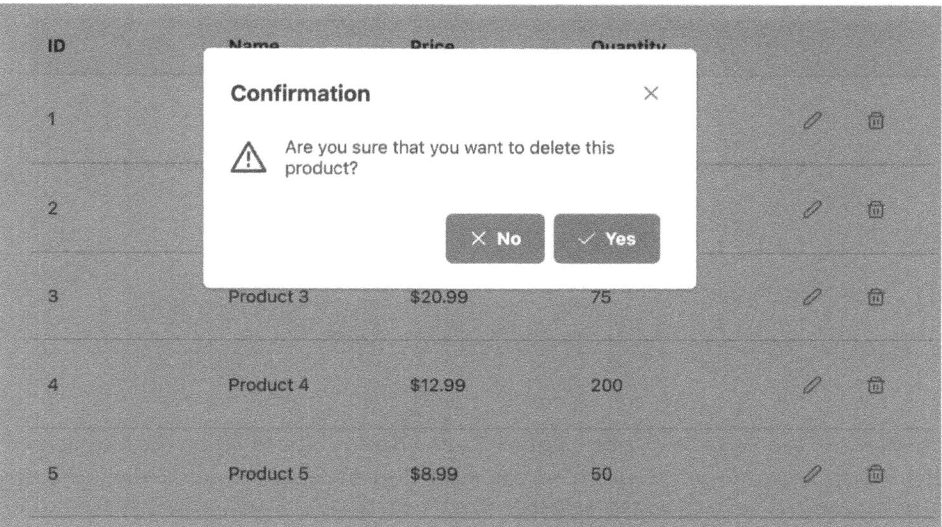

Figure 13.3 – Sample editable table

The table presents a list of products that are editable and removable. Upon clicking the delete icon, a confirmation dialog appears to verify the deletion action.

So, let's now look at some testing tips.

Isolate unit tests

When writing unit tests, it's crucial to isolate the component or service undergoing testing from its dependencies. Angular's **TestBed** provides a powerful toolset for creating and configuring testing modules. By leveraging TestBed, we can mock dependencies and provide fake implementations, enabling us to focus solely on the unit we want to test.

Consider the following example:

```
// tips.component.spec.ts

const productsStub = [
  ...
]
```

```
describe('TipsComponent', () => {
  beforeEach(async () => {
    await TestBed.configureTestingModule({
      imports: [TipsComponent],
      providers: [
        ...
        {
          provide: ShopService,
          useValue: {
            getProducts: jest.fn().mockReturnValue(productsStub),
          },
        },
      ],
    }).compileComponents()
  })
  ..
})
```

In this example, we use `TestBed.configureTestingModule` to set up the testing module. By specifying `imports` and `providers`, we can mock dependencies and ensure they are correctly injected into the component or service being tested. In this case, instead of calling the `getProducts()` function from `ShopService`, we replace the result with the `productsStub` value, which makes it isolated and easier to test.

Utilize NO_ERRORS_SCHEMA

When testing Angular components, we often encounter situations where we don't need to assert the behavior or rendering of child components such as Angular Material components. In such cases, `NO_ERRORS_SCHEMA` can be a handy tool to simplify our test setup.

`NO_ERRORS_SCHEMA` tells Angular's compiler to ignore unrecognized elements and attributes within the component template. This allows us to focus on testing the component's logic without the need to provide detailed mock implementations of child components.

Here's an example:

```
import { NO_ERRORS_SCHEMA } from '@angular/core';

describe('AppComponent', () => {
  let component: AppComponent;
  let fixture: ComponentFixture<AppComponent>;

  beforeEach(async () => {
```

```
    await TestBed.configureTestingModule({
      declarations: [AppComponent],
      schemas: [NO_ERRORS_SCHEMA]
    }).compileComponents();
  });

  // Additional tests go here
});
```

In this example, we specify `schemas: [NO_ERRORS_SCHEMA]` in the testing module configuration. This allows us to test `AppComponent` without worrying about the presence or behavior of any child components.

> **Note**
>
> Avoid using `NO_ERRORS_SCHEMA` if you are writing integration tests. This is because the `NO_ERRORS_SCHEMA` option will ignore template errors for unknown elements and attributes. It allows Angular to run tests even if there are issues with child components' templates. You can read more at `https://angular.io/guide/testing-components-scenarios#no_errors_schema`.

Utilize the spyOn method

Method spying is a powerful technique that allows us to observe and control the behavior of methods during testing. Jest/Jasmine provides the `spyOn` function, along with the spy object, to facilitate method spying.

By using `spyOn`, we can replace a method with a spy function that records all invocations and provides additional capabilities, such as returning specific values or throwing exceptions. This enables us to verify whether a method was called, how many times it was called, and with which arguments.

Consider the following example:

```
// tips.component.spec.ts

beforeEach(async () => {
  await TestBed.configureTestingModule({
    imports: [TipsComponent],
    ...
  }).compileComponents()

  fixture = TestBed.createComponent(TipsComponent)

  confirmDialog = fixture.debugElement.query(
```

```
        By.css('p-confirmdialog')
      ).componentInstance

  })

  it('should show accept message on delete', () => {
    const messageSpy = jest.spyOn(messageService, 'add')

    component.onRowDelete(1)
    fixture.detectChanges()

    confirmDialog.accept()

    expect(messageSpy).toHaveBeenCalledWith({
      severity: 'info',
      summary: 'Confirmed',
      detail: 'Your product is deleted',
    })
  })
```

Let's break down the code in the following list and learn what it does:

- `('should show accept message on delete', () => { ... })`: This is a test case description. In this case, it is checking whether a specific message is shown after deleting a row.
- `const messageSpy = jest.spyOn(messageService, 'add')`: This line creates a spy using Jest's `spyOn` function. It spies on the `add` method of the `messageService` object. This allows us to track whether and how this method is called during the test.
- `component.onRowDelete(1)`: This line calls the `onRowDelete` method of the component under test and passes `1` as an argument. This simulates the deletion of a row with an index of `1`.
- `fixture.detectChanges()`: This line triggers change detection in the test fixture. It ensures that any changes in the component's template are applied and updates are reflected in the test environment.
- `confirmDialog.accept()`: This line simulates the user accepting a confirmation dialog. It assumes that the component has a `confirmDialog` object with an `accept` method, which is called to confirm the deletion.
- `expect(messageSpy).toHaveBeenCalledWith({ ... })`: This line uses the `expect` function to make an assertion. It checks whether the `add` method of the `messageService` object was called with the expected argument. In this case, it expects that the method was called with an object containing specific properties such as `severity`, `summary`, and `detail`.

In conclusion, the spying method is a powerful tool for tracking and controlling function calls during testing, providing insights into the behavior of your application. As we transition to the next section, let's explore how to handle asynchronous code and manage time-related operations in Angular testing.

Working with fakeAsync

`fakeAsync` is a utility in Angular testing that enables synchronous testing of asynchronous code. It runs the test in a special "fake" zone where asynchronous operations can be controlled using `tick()`. Here is a simple demonstration of `fakeAsync` and `tick`:

```
import { fakeAsync, flush, tick } from '@angular/core/testing'

describe('FakeAsync Example', () => {
  it('should test asynchronous code using fakeAsync', fakeAsync(() =>
{
    let value: string | undefined

    // Simulate an asynchronous operation
    setTimeout(() => {
      value = 'completed'
    }, 1000)

    // Use tick to simulate the passage of time
    tick(500) // Simulate 500 milliseconds passed
    expect(value).toBeUndefined() // Value should still be undefined

    tick(500) // Simulate another 500 milliseconds passed
    expect(value).toBe('completed')

    flush()
  }))
})
```

In this example, `fakeAsync` is used to wrap the `test` function. `setTimeout` is used to simulate an asynchronous operation, and `tick` is used to simulate the passage of time. By calling `tick(500)`, you simulate the passing of 500 milliseconds, and then you can make assertions about the state of your application. This is useful for testing asynchronous behavior in a synchronous manner.

Now let's have a look at some more example code that tests the behavior of PrimeNG `MessageService` in `TipsComponent`:

```
// tips.component.spec.ts

import {
```

```
  fakeAsync, flush, tick
} from '@angular/core/testing'

it('should show close message on delete', fakeAsync(() => {
  const messageSpy = jest.spyOn(messageService, 'add')

  component.onRowDelete(1)
  fixture.detectChanges()

  tick(300)

  // Send Escape event
  const escapeEvent: any = document.createEvent('CustomEvent')
  escapeEvent.which = 27
  escapeEvent.initEvent('keydown', true, true)
  document.dispatchEvent(escapeEvent as KeyboardEvent)

  expect(messageSpy).toHaveBeenCalledWith({
    severity: 'warn',
    summary: 'Cancelled',
    detail: 'You have cancelled',
  })

  flush()
}))
```

This test verifies whether the message displays the correct value upon pressing the *Esc* button. The following bullet points outline the flow of the test:

- First, we trigger the delete action through the `component.onRowDelete(1)` function, so the confirmation dialog shows up.
- After that, we use `tick` to advance the virtual clock by a specified amount of time. In this case, we use `tick(300)` to simulate a 300-millisecond delay.
- After the `tick` function is used, we simulate the `Escape` keydown event by creating a `KeyboardEvent` object and dispatching it on the document.
- After the `Escape` event is dispatched, we can assert on the message service via `expect(messageSpy).toHaveBeenCalledWith({...})`.
- Finally, we use `flush` to flush any pending asynchronous tasks, ensuring that all asynchronous operations have been completed before the `test` function exits.

In this section, we've gained insights into managing asynchronous operations in Angular testing. In the next section, we will delve into a powerful testing utility that simplifies and enhances our testing capabilities: Spectator.

Utilizing third-party libraries – Spectator

Spectator (`https://ngneat.github.io/spectator`) provides a set of utilities and techniques that facilitate writing concise and expressive tests for Angular components. It allows you to create component instances, mock dependencies, access the component's DOM, and assert the component's behavior with minimal boilerplate code. It also provides a clean and intuitive syntax that makes test cases more readable and maintainable.

To use it, first, we need to install the `spectator` package by running the following command:

```
npm install @ngneat/spectator --save-dev
```

After that, let's create an example test in which we use `Spectator` for testing an Angular component:

```
// spectator.component.spec.ts

import {
  Spectator,
  createComponentFactory,
  mockProvider,
} from '@ngneat/spectator/jest'

describe('TipsComponent', () => {
  let spectator: Spectator<TipsComponent>

  const createComponent = createComponentFactory({
    component: TipsComponent,
    providers: [
      mockProvider(ShopService, {
        getProducts: () => productsStub,
      }),
    ],
  })

  beforeEach(() => (spectator = createComponent()))

  it('should show table content', () => {
    const table = spectator.query('p-table')
    expect(table?.textContent).toContain('Product 1')
    expect(table?.textContent).toContain('Product 2')
```

```
  })
})
```

The given code utilizes Spectator to test the behavior of the `TipsComponent` instance. Let's break down the code block in the following list so we can understand each part:

- `const createComponent = createComponentFactory({ ... })`: This defines a `createComponent` function using `createComponentFactory` from `Spectator`. This function is used to create an instance of `TipsComponent` for testing. It also provides a mock instance of `ShopService` through the `providers` array.
- `beforeEach(() => (spectator = createComponent()))`: This is a setup step that runs before each test case. It creates a new instance of `TipsComponent` using the `createComponent` function and assigns it to the `spectator` variable.
- `it('should show table content', () => { ... })`: This line defines a test case with the description `should show table content`. This test case verifies that the table in `TipsComponent` contains the expected content. This is broken down in more detail as follows:
    - `const table = spectator.query('p-table')`: This uses the `query` method from the `spectator` object to find the `<p-table>` element in the component's template.
    - `expect(table?.textContent).toContain(...)`: This asserts that `textContent` of the `table` element (if it exists) contains the string `Product 1` or `Product 2`.

As we conclude our exploration of `Spectator`, we've discovered its efficacy in simplifying and improving our testing workflow for Angular components. Now, let's have a look at another helpful library to enhance our testing practices: `ng-mocks`.

Utilizing third-party libraries – ng-mocks

Another powerful third-party library for testing Angular applications is **ng-mocks** (`https://ng-mocks.sudo.eu`). `ng-mocks` simplifies the testing process by providing flexible mocking and stubbing capabilities, making it easier to isolate components and services during testing.

With ng-mocks, we can create mock implementations of Angular components and services, define custom behavior for methods, and verify interactions. This allows us to focus on testing specific units without worrying about the complexities of real implementations.

To use it, first, we need to install the `ng-mocks` package by running the following command:

```
npm install ng-mocks --save-dev
```

Now let's have a look at how to use ng-mocks in Angular testing:

```
// ng-mocks.component.spec.ts

import { MockInstance, MockProvider } from 'ng-mocks'

describe('TipsComponent', () => {
  beforeAll(() =>
    MockInstance(ShopService, () => ({
      getProducts: () => productsStub,
    }))
  )


  beforeEach(async () => {
    await TestBed.configureTestingModule({
      imports: [TipsComponent],
      providers: [
provideNoopAnimations(),
MockProvider(ShopService)
],
    }).compileComponents()

    fixture = TestBed.createComponent(TipsComponent)
    component = fixture.componentInstance

    fixture.detectChanges()
  })
})
```

Let's break down the code block in the following list so that we can understand what it does:

- `providers: [MockProvider(ShopService)]`: This is used to create a mock version of `ShopService` and configure it as a provider in the test environment. This ensures that the test uses the mock version of `ShopService` instead of the actual implementation.
- `beforeAll(() => MockInstance(ShopService, () => ({ ... })))`: This `beforeAll` hook is executed once before all the test cases in the suite. It mocks the `ShopService` dependency by using the `MockInstance` function from `ng-mocks`. It replaces the original `getProducts` method with a mocked implementation that returns `productsStub`.

Note

`ng-mocks` also works fine with `spectator`. You can read more at `https://ng-mocks.sudo.eu/extra/with-3rd-party`.

And with that, we've delved into the intricacies of Angular testing, uncovering practical tips and tricks that empower you to enhance the efficiency and resilience of your test suites.

Summary

In this chapter, you learned how to effectively test PrimeNG components in your Angular applications. By leveraging Jest, a powerful testing framework, you gained the ability to ensure the functionality and reliability of these components.

Throughout this chapter, you explored various concepts and techniques related to testing PrimeNG components. You started by understanding the importance of testing and the benefits it brings to your development workflow. Then, you delved into the specific steps involved in writing unit tests for PrimeNG components, including component setup, testing component behavior, and validating component appearance and interactions.

In addition, you also encountered practical examples, code snippets, and best practices that demonstrated how to effectively test PrimeNG components. By following along and implementing these testing strategies, you gained hands-on experience in verifying the correctness, reliability, and performance of your PrimeNG components.

As we transition to the next chapter, we will shift our focus to the world of building a responsive web application utilizing PrimeNG components.

Part 4: Real-World Application

In this final part, you will apply all the knowledge and skills you have acquired throughout the previous chapters to build a real-world responsive web application. This part will provide you with hands-on experience, allowing you to put your PrimeNG and Angular expertise to practical use.

By the end of this part, you will have completed a fully functional responsive web application using PrimeNG and Angular.

This part includes the following chapter:

- *Chapter 14, Building a Responsive Web Application*

14
Building a Responsive Web Application

In this chapter, you will apply your existing knowledge by building a responsive web application using Angular and PrimeNG components. The focus will be on creating an application that adapts seamlessly to different screen sizes and devices. By the end of this chapter, you will have a solid understanding of how to design and develop responsive web applications using Angular and PrimeNG.

The goal of this chapter is to provide you with the knowledge and skills necessary to build responsive web applications. You will observe how to create a project structure that implements responsive layouts that adapt to different screen sizes and leverage PrimeNG and PrimeFlex to enhance the user experience. Additionally, you will gain insights into deploying the application, ensuring that it reaches a wide audience.

In this chapter, we will cover the following topics:

- Introduction to building a responsive web application
- Introducing our responsive web application project
- Creating the layout of the website
- Starting to develop the website
- Deploying the responsive web application

Technical requirements

This chapter contains code examples for an Angular application. You can find the related source code in the `chapter-14` folder of the following GitHub repository: `https://github.com/PacktPublishing/Next-Level-UI-Development-with-PrimeNG/tree/main/apps/chapter-14`.

Introduction to building a responsive web application

Responsive web applications have become a necessity in today's digital landscape. With users accessing websites and applications on various devices, it is crucial for you to create experiences that seamlessly adapt to different screen sizes and orientations. In this section, we will provide an overview of responsive web applications and explore why they are essential in delivering a satisfying user experience.

Why responsive web applications matter

In the past, websites were primarily designed for desktop computers with fixed screen sizes. However, with the rapid proliferation of smartphones, tablets, and other mobile devices, the landscape of web browsing has dramatically changed. Users now expect websites and applications to be accessible across a wide range of devices, from large desktop monitors to small smartphone screens. Here is an example of a broken website on mobile:

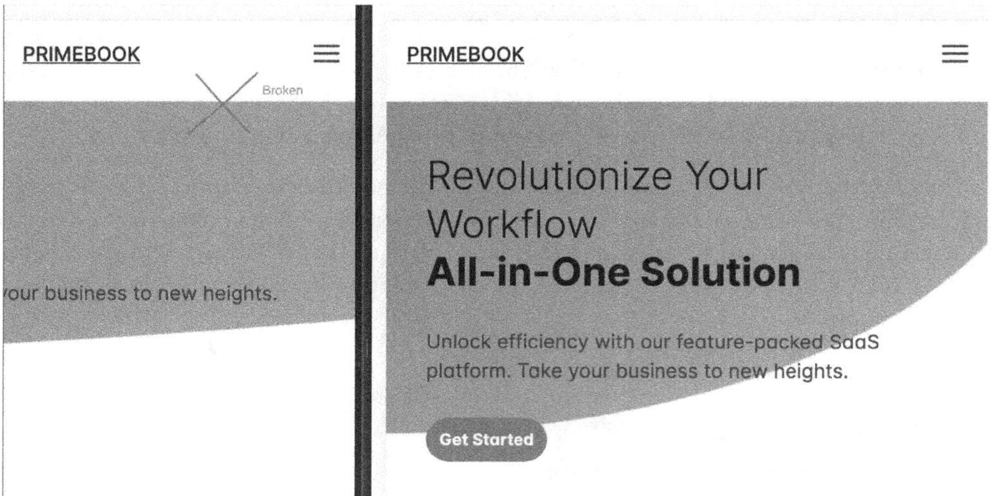

Figure 14.1 – Example of broken layout on mobile

Based on the figure, it is evident that while the website functions properly on larger screens, the layout is broken on smaller devices, rendering the **Get Started** button unresponsive and hindering user interaction.

Responsive web applications are designed to address this problem by dynamically adapting the layout and content based on the user's device and screen size. With a responsive design, you can ensure that the applications are usable and visually appealing across different devices, improving user engagement and satisfaction.

The benefits of responsive web applications

From improved user experience to increased SEO rankings, responsive web applications offer numerous advantages that can greatly enhance your development efforts. Let's review these benefits and understand their impact:

- **Improved user experience**: One of the primary benefits of responsive web applications is providing a consistent and optimized user experience across devices. By adapting the layout, content, and functionality to different screen sizes, users can easily navigate and interact with the application, regardless of the device they are using. This seamless experience enhances user satisfaction and encourages them to engage with the application for longer periods.
- **Increased reach and accessibility**: Responsive web applications have a broader reach as they cater to users on multiple devices. By delivering a consistent experience across different platforms, you can ensure that the applications are accessible to a larger audience. This accessibility is particularly crucial for businesses targeting mobile users, as mobile devices have become the primary means of accessing the internet for many people worldwide.
- **Search engine optimization (SEO) benefits**: Search engines, such as Google, prioritize mobile-friendly websites in the search results. Responsive web applications that provide a seamless user experience across devices are more likely to rank higher in **search engine results pages (SERPs)**. By incorporating responsive design principles, you can improve the application's visibility and attract more organic traffic.

Key principles of responsive web design

Responsive web design follows several key principles to ensure effective adaptation to different screen sizes and devices. Let's see these principles:

- *Fluid grids*: Fluid grids are the foundation of responsive web design. Instead of using fixed pixel-based layouts, you can use proportional units, such as percentages, to define the width and height of elements. This allows content to flexibly adjust and fill the available screen space, creating a fluid and adaptable layout.
- *Flexible images and media*: Images and media play a crucial role in web applications. To ensure they adapt to different screen sizes, you can utilize CSS techniques such as `max-width: 100%` to make images and media elements scale proportionally within the containers. This prevents images from overflowing or being too small on smaller screens.
- *Media queries*: Media queries enable you to apply specific CSS rules based on the characteristics of the user's device, such as screen size, resolution, and orientation. By defining breakpoints at certain screen widths, you can modify the layout, typography, and other elements to optimize the application's appearance on different devices.

- *Mobile-first approach*: The mobile-first approach is a design philosophy that prioritizes designing for mobile devices first, then progressively enhancing the application for larger screens. By starting with a minimalistic and focused design for smaller screens, you can ensure that the core functionality and content are accessible to all users. As the screen size increases, additional features and layout enhancements can be introduced.

Now that you've gained insights into building a responsive web application, let's delve into the specifics of our project. Understanding the fundamentals sets the stage for creating an efficient project structure and crafting a visually appealing and user-friendly website.

Introducing our responsive web application project

Our project focuses on creating a responsive landing page that effectively showcases a **Software as a Service** (**SAAS**) product. The landing page consists of several key sections, including the header, hero, features, testimonials, pricing, and footer. Each section is strategically designed to engage visitors and drive conversions, whether by turning them into customers or capturing them as valuable leads. Let's have a look at the following sketch:

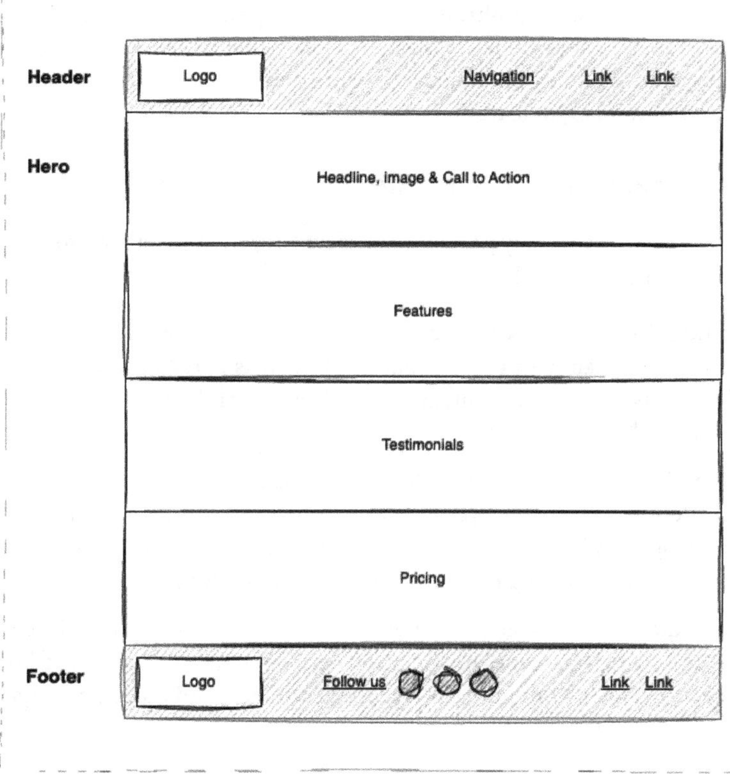

Figure 14.2 – Project initial sketch

Based on the sketch, let's examine the purpose of each section:

- **Header**: The gateway to our application, the header, will feature a clean and intuitive navigation menu, ensuring users can effortlessly explore the various facets of our landing page.
- **Hero**: Our hero section will be a visual delight, instantly grabbing attention with compelling imagery and a succinct message that communicates the essence of our product or service. Moreover, with a strategically placed and persuasive call-to-action button, we will encourage visitors to take the next step, such as signing up, requesting a demo, or exploring our offerings further.
- **Features**: Here, users will delve into the core features that make our offering stand out. Each feature will be elegantly presented, accompanied by captivating visuals and concise descriptions.
- **Testimonials**: Building trust is paramount, and the testimonials section will serve as a testament to the positive experiences of our users. Genuine quotes and perhaps even images will add a personal touch.
- **Pricing**: For users ready to take the next step, the pricing section will provide transparent details about our offerings. We'll strive for clarity and simplicity, making it easy for users to choose the plan that suits them.
- **Footer**: Closing the narrative, the footer will house essential links, contact information, and perhaps a call-to-action button, ensuring users can seamlessly navigate to other parts of our site.

Here is a sneak peek of our final product:

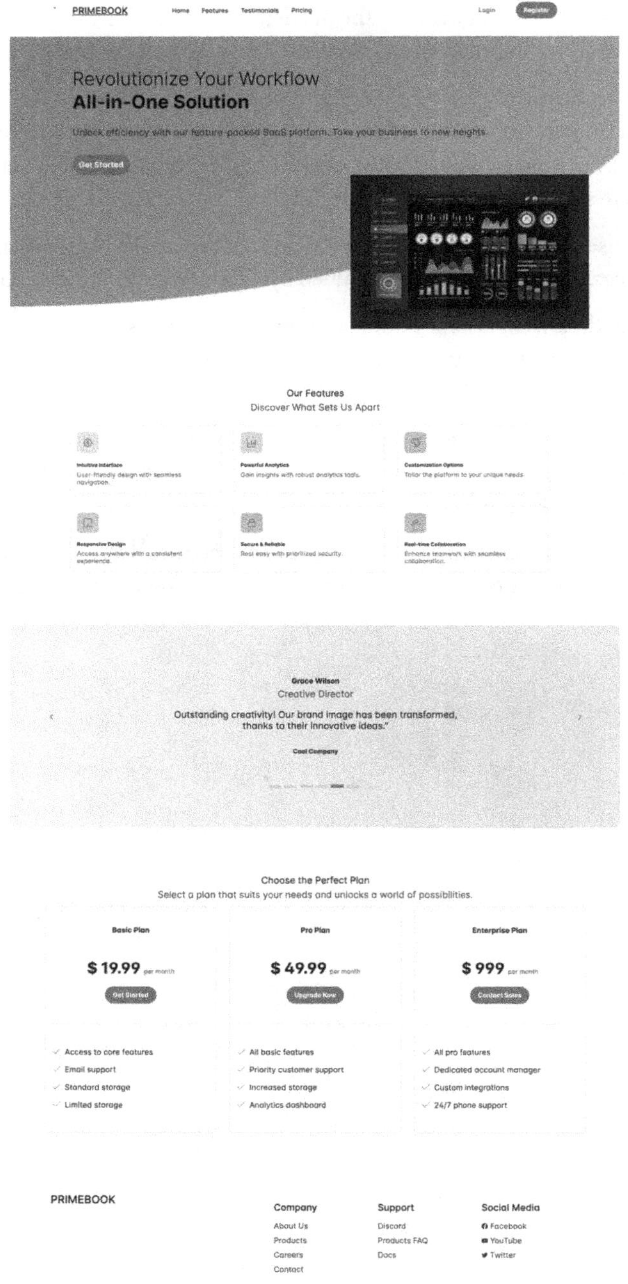

Figure 14.3 – Finalized project

This is the completed product that we're going to build in this chapter. For a better understanding of the code, please check out our GitHub repository for this chapter. Please note that we will mainly focus on the UI, which means these features such as `Login`, `Register`, or `Purchase` will not be implemented.

With a clear project overview in mind, we're ready to transition into the exciting phase of creating the foundational layout for our website.

Creating the layout of the website

Once we have the initial idea for our website, it becomes relatively straightforward to transform those sections into Angular components. Let's have a look at the following transformation:

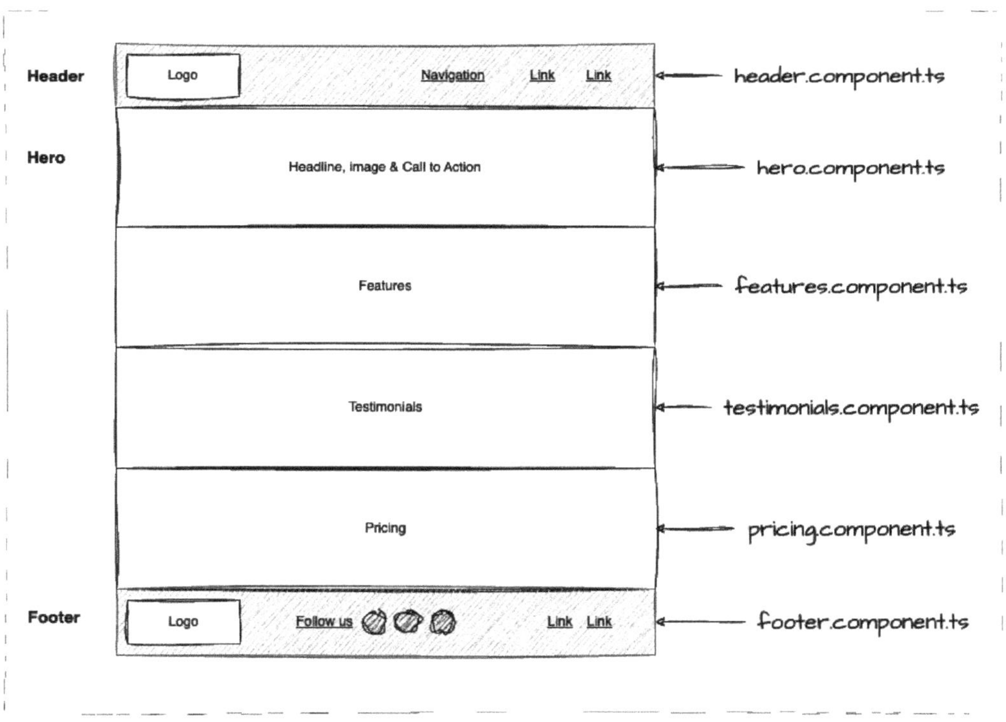

Figure 14.4 – Breaking down sections into Angular components

As depicted in the figure, each section is distinct and can be readily mapped to an Angular component. For instance, the `header` section can be assigned to `header.component.ts`. By following this approach, we can establish the following code structure as an illustrative example:

```
// app.component.ts

<primebook-header />
<primebook-hero />
<primebook-features />
<primebook-testimonials />
<primebook-footer />
```

Each selector represents the corresponding section in our sketch or design.

It's pretty simple to get started to create those components by running the following command:

```
primengbook$ ng g c header --inline-template --inline-style

CREATE src/app/header/header.component.spec.ts (596 bytes)
CREATE src/app/header/header.component.ts (295 bytes)
```

The command will generate a standalone `Header` component with inline styles and an inline template. If you prefer to generate separate files for `header.component.html` and `header.component.scss`, you can omit the `--inline-template` and `--inline-style` options. Once the `Header` component is generated, you can proceed to generate additional components such as `Hero`, `Features`, and more.

Now that we have established the essential layout of our website, it's time to delve into the hands-on development process. We'll be translating the design into code, bringing life to each section.

Starting to develop the website

So far, we've laid out the blueprint for our responsive web application. Now, let's dive into the implementation details of each section. Remember, our goal is to create a seamless and visually appealing experience using Angular, PrimeFlex, PrimeIcons, and PrimeNG components.

Prerequisites

Before getting started, make sure that you have everything configured in your project such as `primeng`, `primeflex`, and `primeicons`. If not, you can run the following command:

```
npm i primeng primeflex primeicons
```

After that, check that your styling and theming are updated. Here is an example of a `styles.scss` file:

```
// src/styles.scss

@import 'primeflex/primeflex.scss';

@import 'primeng/resources/themes/lara-light-blue/theme.css';
@import 'primeng/resources/primeng.css';

@import 'primeicons/primeicons.css';
```

The provided code imports style files from different libraries. By importing these files, you are incorporating the styles, class utilities, themes, and icons provided by `primeflex`, `primeng`, and `primeicons` into your application, enabling you to utilize their features and visual styles.

Header with navigation

The header section plays a vital role in providing users with easy navigation and a seamless user experience. To implement the header, we'll start by creating the HTML structure and applying CSS class utilities. Here's an example of how you can structure the header:

```
// src/app/components/header/header.component.ts

imports: [CommonModule, ButtonModule, RippleModule, StyleClassModule],
...

<header id="header">
  <a class="flex align-items-center" href="#">
    PRIMEBOOK
  </a>
  <a
    class="cursor-pointer block lg:hidden"
    pStyleClass="@next"
    enterFromClass="hidden"
    leaveToClass="hidden"
  >
    <i class="pi pi-bars text-4xl"></i>
  </a>
  <nav
    class="hidden lg:flex absolute lg:static w-full"
    style="top:120px"
  >
    <ul ...>
      <li>
```

```
        <a ...>
          <span>Home</span>
        </a>
      </li>
      ...
    </ul>
  ...
  </nav>
</header>
```

Let's break down the code:

- `<ul ...>...</ul>`: We've used a `<ul>` element to create an unordered list of navigation items. Each item is represented by an `<li>` element, and the corresponding links are wrapped in `<a>` tags. Here is the result of the navigation on desktop:

Figure 14.5 – Navigation layout on desktop

- `<a class="... lg:hidden" pStyleClass="@next" ...>`: This is the line of code that makes the navigation responsive on smaller devices. Let's break it down further:
  - `lg:hidden`: This utilizes `primeflex` utilities, which will hide the hamburger menu on larger screens.
  - `pStyleClass="@next"`: This `StyleClass` feature targets the next element, which is the `nav` element. In this case, when clicking on the hamburger menu on mobile devices, it will show or hide the navigation menu.
- `<nav class="hidden lg:flex ...>`: This represents a navigation menu that is hidden on small screens (`hidden`) and becomes visible on larger screens (`lg:flex`).

Here is the navigation on mobile devices:

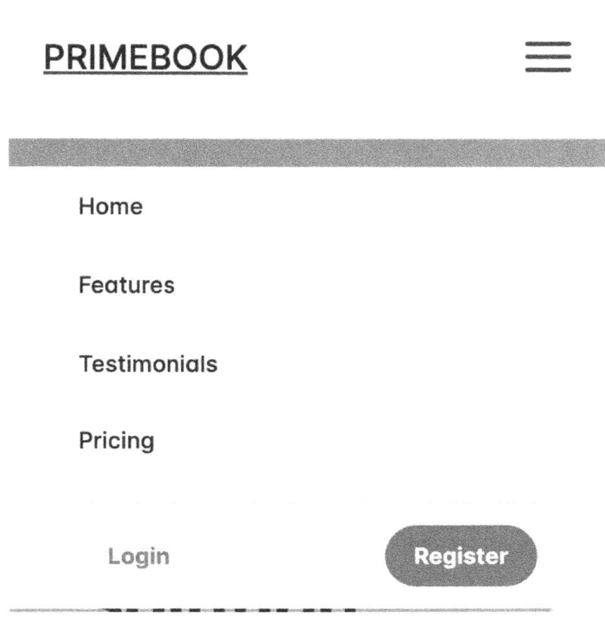

Figure 14.6 – Navigation on mobile devices

In the figure, it is noticeable that instead of displaying the complete navigation menu, only a hamburger icon is shown to toggle the visibility of the navigation menu, which will save space and provide a clean user interface.

Hero section

The hero section is a visually prominent section that immediately captures the attention of visitors, often including a compelling headline, a concise subheading, and a visually appealing image or video. The hero section aims to create a strong first impression and effectively communicate the value proposition of the product or service. Here is the simplified code for our hero section:

```
// src/app/components/hero/hero.component.ts

<section id="hero">
  <div>
    <h1>
      <span class="font-light block">Revolutionize Your Workflow</span
      >All-in-One Solution
    </h1>
```

```
    <p>
      Unlock efficiency with our feature-packed SaaS platform. Take
your
      business to new heights.
    </p>
    <button
      pButton pRipple
      type="button" label="Get Started"
    ></button>
  </div>
  <div class="hidden md:flex">
    <img
      src="assets/hero-image.png"
      alt="Hero Image"
      class="w-9 md:w-auto"
    />
  </div>
</section>
```

Overall, this code snippet represents a hero section with a heading, description, and a call-to-action PrimeNG button. It also includes an image that is displayed on medium-sized (`md:flex`) and larger screens, while hidden on smaller screens (`hidden`).

> **Note**
>
> The following screen size breakpoints are used in PrimeFlex:
>
> - `sm`: Small screens (576 px and above)
>
> - `md`: Medium screens (768 px and above)
>
> - `lg`: Large screens (992 px and above)
>
> - `xl`: Extra-large screens (1,200 px and above)

Let's have a look at the result:

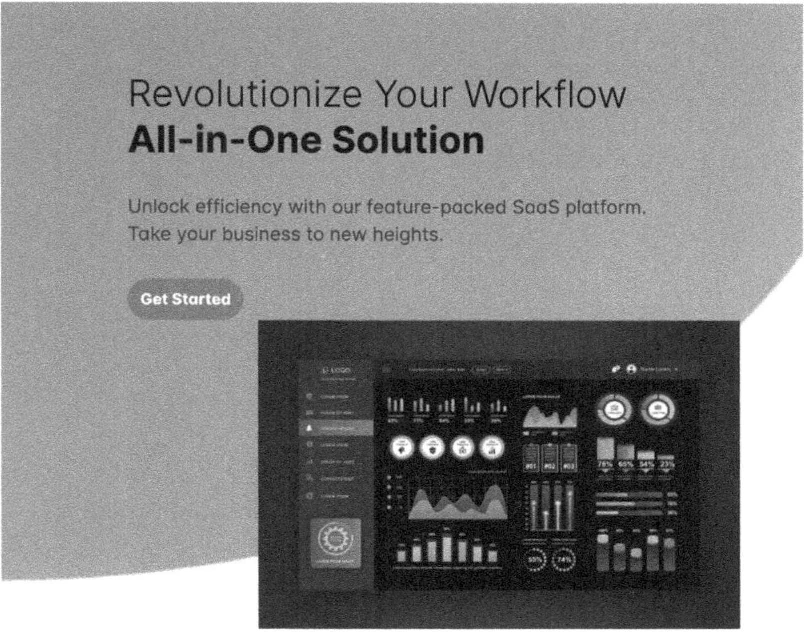

Figure 14.7 – Hero section (credit to Adobe Stock)

Features section

The features section highlights the key features or benefits of the product or service being promoted, aiming to showcase the unique selling points of the product or service and convince visitors of its value. It typically includes a set of visually appealing icons or images, accompanied by a brief description of each feature. Let's have a look at the simplified code for our features section:

```
// features.component.ts

<section id="features">
  <div class="grid justify-content-center">
    <div class="col-12">
      <h2>Our Features</h2>
      <span class="text-600 text-2xl">Discover What Sets Us Apart</
span>
    </div>

    <div
      class="col-12 md:col-12 lg:col-4"
```

```
        *ngFor="let feature of features"
      >

      <div class="{{ feature.bg }}">
        <i class="{{ feature.icon }}"></i>
      </div>
      <h5>{{ feature.heading }}</h5>
      <span>{{ feature.content }}</span>
      </div>
    </div>
</section>

...

features = [
    {
      heading: 'Intuitive Interface',
      content: 'User-friendly design with seamless navigation.',
      icon: 'pi pi-cog',
      bg: 'bg-yellow-200',
    },
    ...
]
```

The provided code represents an HTML structure for a section that showcases a set of features. It also includes an Angular directive (`*ngFor`) to dynamically generate feature elements based on an array of `feature` objects. Let's break it down:

- `<section id="features">`: This line defines a `<section>` element with an `id` attribute set to `"features"`.
- `<div class="grid justify-content-center">`: This code creates a grid-like layout and centers the content horizontally within the grid.
- `<div class="col-12 md:col-12 lg:col-4" *ngFor="let feature of features">`: These classes define the column layout for different screen sizes, which will display three columns on larger screens (`lg:col-4`) and one column on medium and small screens (`col-12`). The `*ngFor` directive is used to iterate over the `features` array, rendering multiple instances of `feature`.
- `features = [...]`: This `features` array contains multiple `feature` objects. Each `feature` object represents an individual feature and includes properties such as `heading`, `content`, `icon`, and `bg`, which will be shown in the template.

> **Note**
>
> The 12-grid system is a widely used framework in web design and layout. It divides the screen into 12 equal columns, providing a flexible and responsive grid structure. This system allows designers and developers to easily create responsive designs by allocating and aligning elements across the grid. We discussed this, along with PrimeFlex, in *Chapter 6*.

Here is the result:

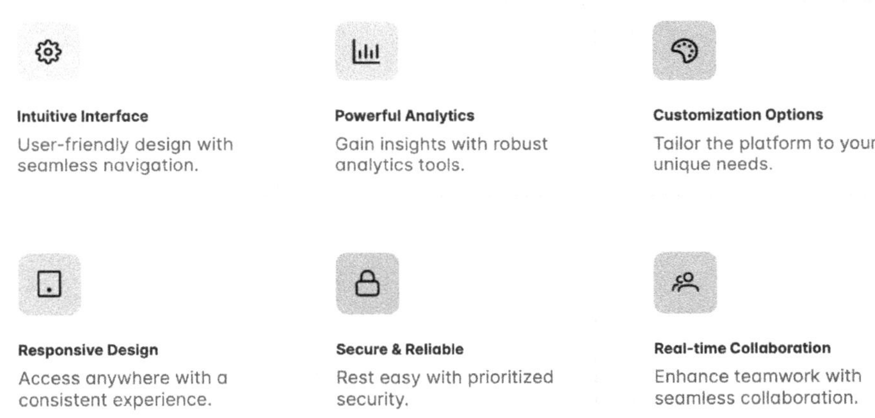

Figure 14.8 – Features section

Testimonials section

The testimonials section provides social proof by showcasing positive feedback or reviews from satisfied customers. It includes testimonials in the form of quotes, customer names, and possibly their profile pictures. This section aims to build trust and credibility, reassuring potential customers that they are making the right choice. Here is the simplified code of our testimonials section:

```
// testimonials.component.ts

<section id="testimonials">
  <p-carousel
    [value]="testimonials"
```

```
    [numVisible]="1"
    [numScroll]="1"
    [circular]="true"
    [autoplayInterval]="3000"
  >
    <ng-template let-testimonial pTemplate="item">
      <h3>{{ testimonial.name }}</h3>
      <span>{{ testimonial.title }}</span>
      <p>
        {{ testimonial.content }}"
      </p>
      <h4>Cool Company</h4>
    </ng-template>
  </p-carousel>
</section>

...

testimonials = [
  {
    name: 'Alice Johnson',
    title: 'CEO',
    content:
      'Exceptional service! The team went above and beyond to meet our
requirements. Highly recommended.',
    company: 'Tech Innovators Inc.',
  },
  ...
]
```

The provided code represents a structure for a testimonial section that utilizes the PrimeNG `Carousel` component. Let's break it down:

- `<p-carousel ...>`: This code will create a carousel and it accepts several input properties:
  - `[value]` is set to the `testimonials` array, which provides the data for the testimonials displayed in the carousel
  - `[numVisible]` is set to `1`, indicating that only one testimonial item should be visible at a time
  - `[numScroll]` is set to `1`, indicating that the carousel should scroll one testimonial item at a time
  - `[circular]` is set to `true`, indicating that the carousel should wrap around and start from the beginning after reaching the end

    - `[autoplayInterval]` is set to `3000`, indicating that the carousel should automatically transition to the next testimonial every 3 seconds

- `<ng-template let-testimonial pTemplate="item">`: This code will set the value of each `testimonial` object to the `testimonial` variable and will be used to render the testimonial detail inside the item template.
- `testimonials = [...]`: This array contains multiple testimonial objects. Each testimonial object represents an individual testimonial and includes properties such as `name`, `title`, `content`, and `company`.

Here is the result:

Figure 14.9 – Testimonials section

Pricing section

The pricing section presents the different pricing options or packages available for the product or service. It typically includes a comparison table, outlining the features and pricing details of each package. The pricing section aims to help potential customers make an informed decision and choose the package that best suits their needs. Here is our simplified code for the pricing table:

```
// pricing.component.ts

<section id="pricing">
  <div class="text-center">
    <h2>Choose the Perfect Plan</h2>
```

```
    <span>
      Select a plan that suits your needs and unlocks a world of
      possibilities.
    </span>
  </div>

  <div class="grid">
    <div class="col-12 lg:col-4" *ngFor="let plan of plans">
      <div>
        <h3>{{ plan.name }}</h3>
        <div>
          <span>$ {{ plan.price }}</span>
          <span>per month</span>
          <button pButton pRipple [label]="plan.cta"></button>
        </div>
        <p-divider class="w-full bg-surface-200" />
        <ul>
          <li *ngFor="let feature of plan.features">
            <i class="pi pi-fw pi-check"></i>
            <span>{{ feature }}</span>
          </li>
        </ul>
      </div>
    </div>
  </div>
</section>

...

plans = [
  {
    name: 'Basic Plan',
    price: 19.99,
    features: [
      'Access to core features',
      'Email support',
      'Standard storage',
      'Limited bandwidth',
    ],
    cta: 'Get Started',
  },
  ...
]
```

The provided code represents a pricing section that displays different plans using a grid layout. Let's break it down:

- `<div class="grid">`: This will create a grid-like layout for our `pricing` section.
- `<div class="col-12 lg:col-4" *ngFor="let plan of plans">`: These classes define the column layout for different screen sizes, which will show one plan on small and medium screens and three pricing plans on larger screens. The `*ngFor` directive is used to iterate over the `plans` array, rendering our pricing table to the DOM.
- `<li ... *ngFor="let feature of plan.features">`: This will loop the features in the plan and render each feature to the list under the unordered list.
- `plans = [...]`: This code represents the `plans` array that contains multiple `plan` objects. Each `plan` object represents an individual plan and includes properties such as `name`, `price`, `features`, and the call-to-action button.

Let's have a look at our pricing section:

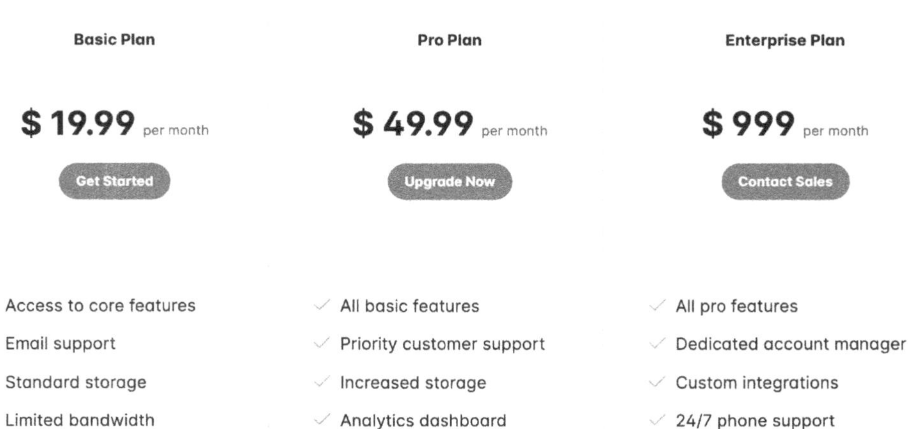

Figure 14.10 – Pricing section

Footer section

The footer section is the bottommost part of the landing page, usually containing additional navigation links, contact information, and copyright notices. It serves as a navigational aid, allowing users to access important links or contact the company for further inquiries. Here is the simplified code for our footer section:

```
// footer.component.ts

<footer>
  <div class="grid justify-content-between">
    <div class="col-12 md:col-2">
      <a>
        <h4>PRIMEBOOK</h4>
      </a>
    </div>

    <div class="col-12 md:col-10 lg:col-7">
      <div class="grid text-center md:text-left">
        <div class="col-12 md:col-4">
          <h4> Company </h4>
          <a>About Us</a>
          ...
        </div>

        <div class="col-12 md:col-4">
          <h4> Support </h4>
          <a>Discord</a>
          ...
        </div>

        <div class="col-12 md:col-4">
          <h4> Social Media </h4>
          <a>
            <i class="pi pi-facebook"></i> Facebook
          </a>
          ...
        </div>
      </div>
    </div>
  </div>
</footer>
```

The provided code represents a footer section. Let's break it down:

- <div class="grid justify-content-between">: This line creates a grid-like layout with items aligned between the start and end of the container.
- <div class="col-12 md:col-2">: This line defines the column layout for different screen sizes for the logo. In this case, it specifies that the column should occupy the entire width for small screens (`col-12`) and 2 columns out of 12 for medium-sized screens (`md:col-2`).
- <div class="col-12 md:col-10 lg:col-7">: This line defines the column layout for different screen sizes for the navigation. In this case, it specifies that the column should occupy the entire width for small screens (`col-12`), 10 columns out of 12 for medium-sized screens (`md:col-10`), and 7 columns out of 12 for large screens (`lg:col-7`). Let's also look into the navigation detail of the footer:
  - <div class="grid text-center md:text-left">: This code creates a grid-like layout with text alignment centered for small screens (`text-center`) and left-aligned for medium-sized screens (`md:text-left`).
  - <div class="col-12 md:col-4">: This line defines the layout for each navigation section. In small screens (`col-12`), the navigation will be full width; however, on medium to large-sized screens, the navigation will be divided into three sections (`md:col-4`).

Let's have a look at the result:

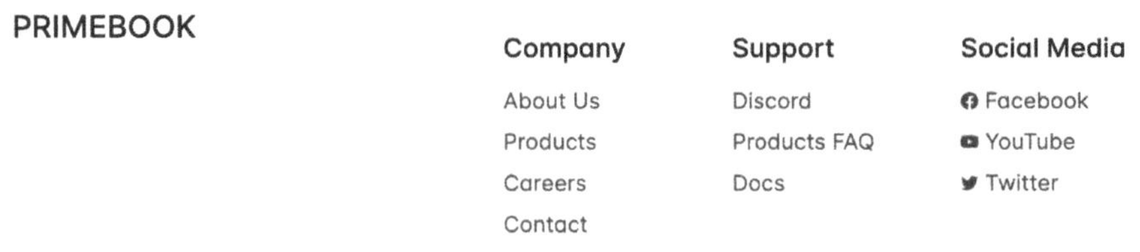

PRIMEBOOK

Company	Support	Social Media
About Us	Discord	Facebook
Products	Products FAQ	YouTube
Careers	Docs	Twitter
Contact		

Figure 14.11 – Footer section

As we conclude the development phase, our responsive web application is now ready for its live deployment. Let's embark on the final step of our journey: the production deployment process.

Deploying the responsive web application

Congratulations! You've successfully built a responsive web application using Angular, PrimeFlex, PrimeIcons, and PrimeNG components. Now, it's time to share your creation with the world. Deploying a web application involves several crucial steps, from preparing the project for production to choosing the right deployment platform. Let's explore each of these aspects to ensure a smooth deployment process.

Getting the project ready for production

Before deploying your web application, it's essential to ensure that everything is optimized and ready for production. Here are a few key steps to follow:

- *Optimize assets*: Make sure to optimize your images to reduce their file size and improve loading times. You can use tools such as image compressors to achieve this. Here is an example of how to serve different images in different screen sizes:

  ```
  <picture>
    <source srcset="image-small.jpg" media="(max-width: 576px)">
    <source srcset="image-medium.jpg" media="(max-width: 992px)">
    <source srcset="image-large.jpg" media="(min-width: 993px)">
    <img src="image-large.jpg" alt="Example image">
  </picture>
  ```

  In the code, the `<picture>` element is used to define a container for multiple image sources. Inside the `<picture>` element, the `<source>` elements are used to specify different image sources based on the media query conditions. For example, the `image-small.jpg` image source will be served when the screen width is at a maximum of 576 pixels.

- *Configure environment variables*: If your application relies on environment variables, ensure that they are properly configured for the production environment. This may include API keys, database connection strings, or other sensitive information.
- *Set up error logging*: Implement error logging to capture and track any runtime errors that may occur in the production environment. Tools such as Sentry or Rollbar can help you track and diagnose issues effectively.

By following these steps, you can ensure that your web application is optimized and ready for deployment.

After that, you can run the following command to build your application:

```
> ng build

Initial Chunk Files    | Names          |  Raw Size | Estimated Transfer
Size
styles-QKTSICQI.css    | styles         | 497.28 kB
|                 33.26 kB
main-A7MWZFB4.js       | main           | 343.22 kB
|                 88.63 kB
polyfills-LZBJRJJE.js  | polyfills      |  32.69 kB
|                 10.59 kB

                       | Initial Total  | 873.20 kB
|                132.48 kB

Application bundle generation complete. [11.871 seconds]
```

You can see that after running the build, we created 3 chunk files with a total time of 11.871 seconds.

If you are using Angular 17, your compiled files are likely under `dist/chapter-14/browser`, like so:

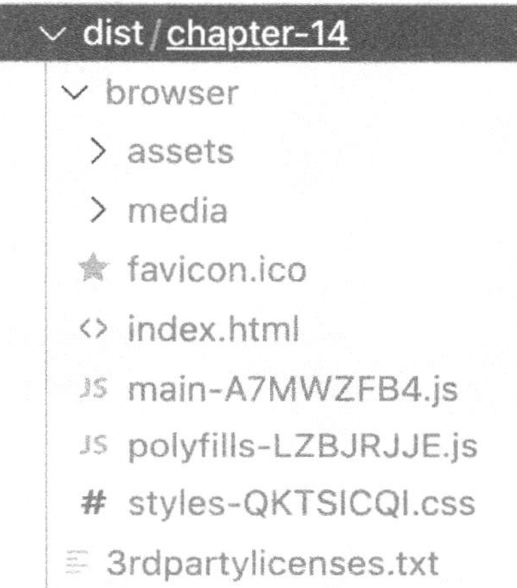

Figure 14.12 – Compiled project folder

For an Angular version less than 17, it will be under `dist/chapter-14`.

Different options for deployment

Once your project is production-ready, the next step is to choose a deployment platform. Several platforms offer seamless deployment for Angular applications. Let's explore a few popular options:

- *Firebase Hosting*: Firebase Hosting (`https://firebase.google.com/docs/hosting`) provides a fast and secure way to host your web application. It supports continuous deployment, SSL, and custom domains. Here is how we start to initialize and deploy the project to Firebase:

```
# Install Firebase CLI
npm install -g firebase-tools

# Set up a project directory
firebase init

# Deploy to Firebase
firebase deploy
```

- *Vercel*: Vercel (`https://vercel.com`) offers a zero-configuration platform for deploying frontend applications. It integrates with your version control system for automatic deployments. Here is how we deploy the project using the Vercel CLI:

```
# Install Vercel CLI
npm install -g vercel

# Deploy to Vercel
vercel
```

- *Netlify*: Netlify (`https://www.netlify.com`) is a powerful platform that automates your build and deployment pipeline. It supports continuous integration and deploys directly from your Git repository. Here is an example of deploying to Netlify from your local machine:

```
# Netlify CLI (if needed)
npm install -g netlify-cli

# Deploy to Netlify
netlify deploy --prod
```

- *GitHub Pages*: If your code is hosted on GitHub, GitHub Pages is a straightforward option. It automatically deploys your app when you push changes to the `gh-pages` branch. You can learn more at `https://docs.github.com/en/pages/getting-started-with-github-pages`.

These are suggested hosting sites where you can deploy your Angular application, but you are free to deploy it anywhere you prefer. If you have your own hosting provider, simply upload the contents of the compiled files to your hosting provider, and it will serve your site seamlessly.

Final notes after deployment

Congratulations, your web application is live! Before celebrating, consider these final notes:

- *Monitor performance*: Regularly monitor your application's performance using tools such as Google Lighthouse or WebPageTest. Address any issues that may impact user experience.
- *Security considerations*: Ensure that your deployed application follows best practices for web security. Use HTTPS, keep dependencies updated, and implement secure authentication mechanisms if applicable.
- *User feedback*: Encourage users to provide feedback on your application. Monitor user reviews, comments, and error reports to continuously improve the user experience.
- *Documentation*: Update your project's documentation to reflect any changes made during deployment. Include information on how others can contribute or report issues.

- *Stay updated*: Keep an eye on updates to Angular, PrimeNG, and any third-party libraries you're using. Regularly update your dependencies to benefit from new features and security patches.

In conclusion, deploying a web application involves careful preparation and the selection of an appropriate platform. By following best practices and choosing a reliable deployment option, you can ensure that your responsive web application reaches its audience seamlessly.

Summary

Congratulations on completing the journey of building a responsive web application with Angular and PrimeNG components! In this final chapter, you've acquired essential insights into creating an efficient project structure, implementing responsive layouts, integrating various PrimeNG elements, and deploying your application to share it with the world.

Throughout the chapter, we've learned that building responsive web applications is crucial in today's digital landscape. Users access applications from various devices and responsiveness ensures a consistent and enjoyable user experience across desktops, tablets, and mobile devices. The knowledge gained in this chapter empowers you to create applications that adapt to different screen sizes, providing accessibility to a broader audience.

As you move forward, the journey offers opportunities for growth and exploration. Continuously enhance your application, stay updated on Angular and PrimeNG releases, and explore advanced topics such as PWAs and server-side rendering. Consider contributing to the open source community and expanding your skills into related technologies. Embrace the path of continuous learning and innovation in web development.

In conclusion, building a responsive web application is a rewarding endeavor that opens doors to a myriad of possibilities. Whether you're creating projects for clients, users, or personal satisfaction, the knowledge gained in this book equips you with the tools to deliver exceptional user experiences. As you move forward, embrace the journey of continuous learning and innovation in the dynamic field of web development. Happy coding!

Index

Symbols

A

O

P

R

U

V

Other Books You May Enjoy

If you enjoyed this book, you may be interested in these other books by Packt:

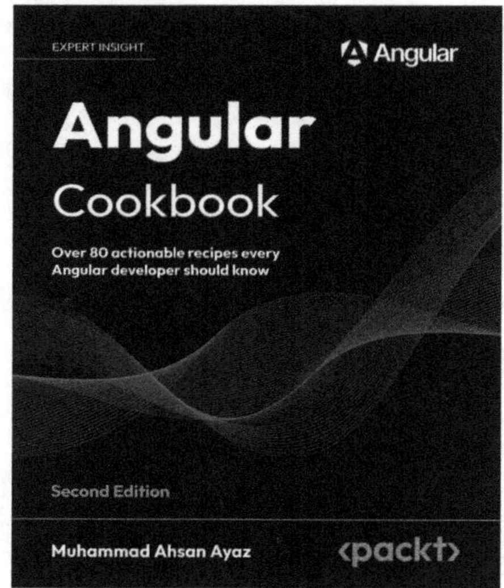

Angular Cookbook

Muhammad Ahsan Ayaz

ISBN: 978-1-80323-344-4

- Gain a better understanding of how components, services, and directives work in Angular
- Get to grips with creating Progressive Web Apps using Angular from scratch
- Build rich animations and add them to your Angular apps
- Manage your app's data reactivity using RxJS
- Implement state management for your Angular apps with NgRx
- Optimize the performance of your new and existing web apps
- Write fail-safe unit tests and end-to-end tests for your web apps using Jest and Cypress
- Get familiar with Angular CDK components for designing effective Angular components

Angular for Enterprise Applications

Doguhan Uluca

ISBN: 978-1-80512-712-3

- Best practices for architecting and leading enterprise projects
- Minimalist, value-first approach to delivering web apps
- How standalone components, services, providers, modules, lazy loading, and directives work in Angular
- Manage your app's data reactivity using Signals or RxJS
- State management for your Angular apps with NgRx
- Angular ecosystem to build and deliver enterprise applications

Packt is searching for authors like you

If you're interested in becoming an author for Packt, please visit `authors.packtpub.com` and apply today. We have worked with thousands of developers and tech professionals, just like you, to help them share their insight with the global tech community. You can make a general application, apply for a specific hot topic that we are recruiting an author for, or submit your own idea.

Share Your Thoughts

Now you've finished *Next-Level UI Development with PrimeNG*, we'd love to hear your thoughts! Scan the QR code below to go straight to the Amazon review page for this book and share your feedback or leave a review on the site that you purchased it from.

`https://packt.link/r/1-803-24981-1`

Your review is important to us and the tech community and will help us make sure we're delivering excellent quality content.

Download a free PDF copy of this book

Thanks for purchasing this book!

Do you like to read on the go but are unable to carry your print books everywhere?

Is your eBook purchase not compatible with the device of your choice?

Don't worry, now with every Packt book you get a DRM-free PDF version of that book at no cost.

Read anywhere, any place, on any device. Search, copy, and paste code from your favorite technical books directly into your application.

The perks don't stop there, you can get exclusive access to discounts, newsletters, and great free content in your inbox daily

Follow these simple steps to get the benefits:

1. Scan the QR code or visit the link below

`https://packt.link/free-ebook/9781803249810`

2. Submit your proof of purchase
3. That's it! We'll send your free PDF and other benefits to your email directly

www.ingramcontent.com/pod-product-compliance
Lightning Source LLC
Chambersburg PA
CBHW080616060326
40690CB00021B/4715
* 9 7 8 1 8 0 3 2 4 9 8 1 0 *